A Native Hawaiian Garden

A Native Hawaiian Garden

How to Grow and Care for Island Plants

John L. Culliney
Bruce P. Koebele

A Latitude 20 Book
University of Hawai'i Press
Honolulu

© 1999 University of Hawai'i Press
All rights reserved
Printed in China
04 5 4 3 2

Library of Congress Cataloging-in-Publication Data

Culliney, John L., 1942–
    A native Hawaiian garden: how to grow and care for island plants
/ John L. Culliney and Bruce P. Koebele.
      p.   cm.
    Includes bibliographical references.
    ISBN 0–8248–2176–9 (pbk. : alk. paper)
    1. Native plant gardening—Hawaii.   2. Native plants for
cultivation—Hawaii.   3. Plants, Ornamental—Hawaii.   I. Koebele,
Bruce, P., 1956–   .   II. Title.
SB439.24.I13C85   1999
635.9'51969—dc21   99–15190
                                                                          CIP

University of Hawai'i Press books are printed on acid-free paper and meet the guidelines for permanence and durability of the Council on Library Resources.

Design by Nighthawk Design

Printed by Toppan Printing Co., (H.K.) Ltd.

Photograph Credits
John L. Culliney—pp. 5, 19, 73a, 133
Priscilla S. Millen—pp. 13, 56b, 71b
Karen E. Shigematsu—pp. 11, 115b
All other photographs by Bruce P. Koebele

*This book is dedicated by John L. Culliney to my daughter
Susan who helped raise plants on Rocky Hill,
and by Bruce P. Koebele to Ralph and Marina Koebele,
whose support and enthusiasm for all my
endeavors have been limitless*

Contents

Preface	ix
Acknowledgments	xi
INTRODUCTION *Our Unique Botanical Heritage*	1
HOW TO GROW HAWAIIAN PLANTS *From Seed to Planting Out*	12
PROPAGATION AND CARE OF SELECTED SPECIES *From 'a'ali'i to wiliwili*	34
LANDSCAPE GUIDE	142
APPENDIX 1 *Associations, Institutions, and Societies Devoted to Native Hawaiian Plants*	149
APPENDIX 2 *Public and Private Gardens and Sanctuaries Featuring Native Hawaiian Plants*	151
Glossary	155
Literature Cited	157
Index	161

Preface

A Native Hawaiian Garden grew out of our love and concern for Hawai'i's native plants. Many of the plants and animals of our Islands have already become extinct as a result of centuries of human impact. Those *kama'āina* organisms that still survive should be thought of as precious antiques wrought by nature over millennia of evolutionary artistry. We believe that Hawai'i's people are gaining a new awareness of the unique living nature that evolved after reaching the Islands against all odds. While a number of "nature books" have been written on the subject of native plants, what we provide here is a substantial guide for people to make direct and lasting contact with some of the rarest life-forms on earth. We hope that people will use our book as they get their hands dirty, develop a Hawaiian green thumb, and become involved literally in saving some of our native plants from extinction.

This book aims at a wide readership. We are confident that *A Native Hawaiian Garden* will be of use not only to gardeners and landscapers but also to professional horticulturists and botanists. We believe our methods for raising Hawaiian plants can be used for large-scale revegetation work or simply restoring a patch of *kama'āina* greenery to a residential backyard.

We provide more detail about individual species—focusing on specific, well-tested methods for cultivating them and problems encountered in their propagation—than any other publication. Unlike other texts, we include a wide variety of trees, for we believe that trees provide an essential framework in most native Hawaiian landscapes, wild or domesticated. They enhance ecological and aesthetic niches of compatible shrubs and other smaller plants; they protect smaller species from strong wind, sun, and desiccation. Further, our emphasis is on plants adapted to dry and mesic (moderate rainfall) regions. These are the climates where most of us live in the Islands and the regions that have suffered the greatest loss of native flora.

Our book covers sixty-three species of native plants, many of which are rare. Reliable propagation techniques have never been described for many of these species. We provide new and extensive guidelines on handling seeds and seedlings, as well as repotting larger plants and establishing them in the ground. Our list of plants, with notes on their history, biology, and horticultural requirements and preferences, includes endemic species that evolved in Hawai'i and occur naturally nowhere else in the world. We also feature indigenous plants—those native to Hawai'i but also found naturally elsewhere, chiefly on other tropical Pacific islands.

One of our chief reasons for writing this book is that Hawai'i has become an alien world. Although our landscapes often appear richly diverse and lush with tropical vegetation, we are surrounded by plants from nearly everywhere else in the world except Hawai'i. We have largely ignored the husbandry of our native plants, even though many are as beautiful and unusual as the aliens. Part of the problem has been the lack of a reliable guide for the gardening public. This book is for people who would like to get to know some of our *kama'āina* plants by raising them in their yards, on farms and ranches, on the grounds of hotels, schools, and on many other landscapes in our communities.

There is another more urgent purpose for this book: Most native Hawaiian plants are in peril. Today, the impacts on native ecosystems are overwhelming. Urbanization, diversion and depletion of natural waters, widespread planting of aggressive alien vegetation, especially grasses, and the introduction of devastating pests are among the factors that have combined to reduce the native flora to its present sad state. This is why it is so important to protect and sustain our remaining native plants. The surviving Hawaiian species are a vital treasure of great scientific interest, of great value in the world's heritage of biodiversity, of significant value to Hawaiian culture, and of unique beauty, graceful presence, and *kama'āina* standing on our midocean landscapes. Symbolically, the beleaguered native flora of Hawai'i looms large. The ecological crisis and destruction of endemic nature in Hawai'i may foretell the fate of our planet itself, for it too is an island of profound isolation.

Acknowledgments

Numerous individuals helped in the germination and development of this book. We especially thank Priscilla Millen, Professor of Botany at Leeward Community College on Oʻahu, whose inspiration, sage advice, and outstanding knowledge of Hawaiʻi's native plants influenced the book from its earliest stage. Professor Millen wrote original drafts for Appendixes 1 and 2 and consulted with the authors on the rest of the manuscript and photographic selections for the book. Among other specialists who guided the authors' efforts, Dr. Thomas Culliney, entomologist for the State of Hawaiʻi, identified insect enemies of the plants we have raised and suggested appropriate pest control measures. Karen Shigematsu, research associate at Harold L. Lyon Arboretum, helped with the editing of Appendixes 1 and 2, provided us with numerous early references on growing native plants, and, in reviewing parts of the manuscript, corrected our sometimes frequent botanical errors. Aaron Lowe, environmental specialist for the State Department of Land and Natural Resources, during the course of several memorable hikes, shared his knowledge of wild native plant communities and natural habitats of the Islands.

Each of the coauthors independently has received help, information, and encouragement from various scientists, planters, and explorers of Hawaiʻi's unique botanical landscape. John thanks Keith Zeilinger, a craftsman in fine woods, for sharing many occasions to learn about native trees and their woodworking properties. Dr. Steve Montgomery influenced the text with scientific information and his strong conservation ethic for Hawaiʻi. Sean Casey, Dan Sailer, and Barbara Kelly of Youth for Environmental Service (YES) developed several planting projects on Oʻahu and coordinated the participation of young people in others, such as the Hanauma Bay Nature Preserve, from which much was learned about growth and survival of native species. The work of YES offers new hope for the future of native landscape restoration in the Islands. John also thanks Barbara, Aaron, and Susan Culliney for all their help, inspiration, and encouragement, and in particular for their patience and tolerance of hundreds of potted plants occupying a tiny backyard.

Bruce thanks Judy and Will Hancock of Kohala, Hawaiʻi, for agreeing to let us use them as a specific example of the success private individuals, with modest resources, can have in rescuing a rare Hawaiian plant *(koaiʻa)* from extinction. Peter Van Dyke, manager at Amy B. H. Greenwell Ethnobotanical Garden, graciously spent an afternoon guiding the author around the garden for many of the photographs in this book. Thanks go also to Liz Huppman, horticulturist at Lyon Arboretum, for her patient answers to frequent questions concerning Hawaiian hibiscus and other native species. Bruce would also like to thank Eric Enos, director of Ka'ala Farm, Inc., for letting the authors use the Ka'ala office to do some final editing of the book.

Both of us are grateful to several institutions and their staffs for a wide range of help and support, from guided tours to donated seeds and sharing of information both spoken and published. These organizations include: Lyon Arboretum, Amy B. H. Greenwell Ethnobotanical Garden, Honolulu Botanical Gardens, Hanauma Bay Nature Preserve, Hawai'i Volcanoes National Park, Bernice P. Bishop Museum, The Nature Conservancy of Hawai'i, Department of Land and Natural Resources (State of Hawai'i), and the U.S. Fish and Wildlife Service.

In writing the manuscript and assembling the photographs, our work has been expertly nurtured and gracefully shaped by Pamela Kelley, our editor at the University of Hawai'i Press. We are grateful for her editorial green thumb. We also wish to acknowledge the constructive comments of two anonymous reviewers. Any errors that remain are our responsibility.

INTRODUCTION

Origins of Hawai'i's Unique Flora

Hawaiian native plants are among the world's most improbable products of evolution. Only by remote chance did the ancestors of our Hawaiian plants, as seeds or spores, make a great transoceanic leap to reach these islands. Even after arriving, they usually faced great odds against their survival. At the very least, each seed and seedling needed a niche in the ground to nurture its growth. For some species a rock crevice might suffice, but a given seed must reach that crevice or perish. There also had to be enough—but not too much—moisture and sunlight. Another vital factor was a proper temperature range for growth and reproduction. Even after sprouting, survival was dependent on the vagaries of volcanic eruptions, hurricanes, tsunamis, landslides, and perhaps hungry plant-eating insects or birds. It is not surprising, therefore, that evolutionary biologists have determined that Hawai'i's major native flora (flowering plants and ferns) arose from only about 386 colonizing ancestors (Sohmer and Gustafson 1987).

The prevailing scientific view is that the founders of the Hawaiian flora were primarily hardy species. Many had small seeds or (in ferns) tiny spores. At least some of these could have been carried by winds such as the prevailing high-altitude jet stream from Asia. Large-volume air samples collected at high altitudes often contain such seeds and spores, as well as microorganisms and even small insects and spiders at altitudes reaching 30,000 feet or more (Gressitt and Yoshimoto 1964). Still others among the colonizers came to Hawai'i as seeds attached to the feathers or in dried mud on the feet of our common migratory birds, such as the wandering tattler and pintail duck. Both of those species, among others, annually fly to Hawai'i from the Pacific Rim. Some seeds are swallowed by birds and pass intact through the avian digestive system to germinate in the bird droppings. The two- to three-day nonstop flights of birds to Hawai'i from North America, Asia, and South Pacific islands would accommodate this gut transport of seeds by plovers and related species (Proctor 1968). Perhaps a few kinds of plants reached the Islands by drifting across the ocean, but this is probably the least likely means for most plants, whose seeds quickly die in salt water.

After they successfully established themselves in healthy populations on a given island, the new arrivals began to spread into the mosaic of Hawaiian habitats on coastal plains, in valleys, and at different elevations on mountain slopes. Occasionally, various established species would cross ocean channels to reach neighbor islands. Through the accumulation of mutations and the process of adaptation, governed by natural selection in different island environments, native Hawaiian plants diversified. Approximately a thousand flowering species and 145 ferns are recognized today by modern botanists (Sohmer and Gustafson 1987).

Some of Hawai'i's more unusual products of evolution are shrubs and trees that have evolved from small, soft-stemmed herbaceous plants. Among the species described in this book that have followed this path, called arborescence, as they evolved in the Islands are *āheahea (Chenopodium),* '*akoko (Chamaesyce),* and '*aiea (Nothocestrum),* a tree nightshade related to tomato and eggplant. Likewise, the *kuluʻī (Nototrichium)* is an amaranth shrub, sometimes becoming a small tree. Its closest relatives are small-stemmed, spreading herbs from Australia (Carlquist 1980).

Another evolutionary change seen in plants as they adapt to island environments with few competing species is loss of dispersability. Their seeds may lose the capacity to cling to birds or other agents of dispersal. They may become heavy so that they sink in seawater or grow too large to be carried by the wind. Hawaiian plants exemplify all of these changes. For example, the *koʻokoʻolau* (several native species of *Bidens*—see our account of raising *Bidens* in a later section) has close alien relatives with barblike extensions of the fruit (achene) that fasten tenaciously to birds' feathers, mammals' fur, and even human clothing. All of the Hawaiian species have lost these barbs, often along with other features such as needlelike shape and hairy coatings that aid dispersal in the nonnative types. Another example is provided by the Hawaiian *wiliwili (Erythrina sandwicensis),* whose seeds sink in seawater, while those of several related nonnative species float. Carlquist (1980) discusses these and several other examples of this phenomenon in Hawaiian plants.

Most Hawaiian plants lack defensive adaptations such as thorns and poisonous sap that develop through natural selection to protect against herbivorous animals such as cattle and goats. An exception to this trend in Hawai'i's flora is reported for *ākia* shrubs *(Wikstroemia),* of which more than a dozen species evolved in Hawai'i. At least some Hawaiian *ākia* may be highly toxic (Stone and Pratt 1994). Of the few native plants that do develop significantly thorny surfaces, at least some are believed to be relatively recent arrivals in the Islands. One example is the *pua kala,* or prickly poppy *(Argemone).* Carlquist (1980) calls it "probably a very recent, barely prehuman arrival in the Hawaiian flora."

Hawai'i's Plants in Peril

Now the native Hawaiian flora is rapidly diminishing. This sad process has been accelerating over the last fifteen hundred years or so since the arrival of human beings

Introduction

Pua kala is one of the few Hawaiian native plants with any defense against large introduced herbivores such as cows and goats. But this variety of *pua kala* from the Big Island, unlike other varieties, no longer has prickles on its stems or fruits.

in the Islands. The native Hawaiians who voyaged from the South Pacific (Marquesas and Society Islands) were pioneers colonizing virgin landscapes. Like all pioneers, they transformed the land in order to build a lasting society. Over approximately ten to twelve centuries, the aboriginal Hawaiian forests were burned and cleared in large areas of the main islands, commonly from the shoreline to 1,200 meters in elevation and sometimes higher (Kirch 1985). Beginning with Captain Cook himself, early Western explorers reported that extensive regions had been cleared of forests (Price 1971). Much of this land, encompassing hundreds of square miles on all the major islands, was put into agriculture and used to sustain long-term population centers. Notable exceptions were on Kahoʻolawe, Lānaʻi, and south Molokaʻi where Hawaiians, after a few generations, abandoned their settlements as the land became desiccated and barren—probably a consequence of deforestation (Culliney 1988). From a naturalist's point of view, much was lost. Hawaiian biodiversity began to decline significantly during this period, and numerous plants and animals in the lowlands became extinct or retreated into less ideal habitats at higher elevations (Olson and James 1982).

The first Hawaiians also introduced rats and pigs to the Islands. Both of these animals have proved potent destroyers of native plants. Rats eat the flowers, seeds, and seedlings of a wide variety of species. Pigs are not only consumers of many specific plants but cause indiscriminate destruction by rooting up swaths of ground.

Decline and extinction of the native flora intensified after the arrival of Captain Cook. In an earlier book, coauthor Culliney categorized the two major onslaughts of ecological change that ensued in the first century after Western contact:

One was an extension of the Polynesian planters' activities, which, as the stone adz gave way to the steel ax, began to reach ever higher into the mountain forests. Soon, industrial-scale deforestation was rampant; whole mountainsides of koa were consumed by sawmills, and other areas clearcut for firewood as waves of sugar cane began to lap up over the mesic slopes. The second major land-transforming force was at least as powerful, even more widespread, and for a long time essentially uncontrollable: the grazing of millions of feral mammals. Except for the uncertain but perhaps light impact of the Polynesian pig, it was unprecedented in the islands (Culliney 1988).

Further changes caused accelerating destruction of Hawai'i's flora and fauna into and throughout much of the twentieth century. Today the onslaught continues and focuses ever more intensely on the remaining small fragments of our lowland Hawaiian plant communities. Several of the ongoing environmental impacts may be more powerful than ever.

In Hawai'i, one of the most destructive factors is fire. Each year, wildfires on various islands consume ever-dwindling populations of our rarest flora. For example, in North Kona on the Big Island, over the past decade, one of the state's last significant stands of native lowland dry forest has lost thousands of acres to several fires. Hundreds of Hawai'i's rarest trees and shrubs, including *kauila, uhiuhi, 'aiea, 'ohe makai,* and others—a major proportion of the wild inventory of these species in the entire archipelago—have been destroyed together with many of the more abundant species: *'ōhi'a lehua, lama, hala pepe, āulu, 'iliahi, wiliwili,* and so on. Throughout the Islands, this is the ecosystem that many botanists believe once hosted the highest vegetational diversity (Carlquist 1980). After each fire, one looks in vain for renewal, but the trees are not growing back.

The burning of the remnant native lowland forest is only the first part of the story. Much of the the land in question is owned by the State of Hawai'i or large private organizations such as the Bishop Estate, but it has been leased to cattle ranchers, in some cases for more than a century. The remaining patches of native forest in these regions are confined to the least accessible lava fields and steep-sided gulches, places the "hooved locusts" could not easily reach. Now, however, the cattle are nearly gone; ranching is a dying business—but one of its legacies is ubiquitous. Alien grasses up to 2 meters high sprout from thick tussocks and in many areas blanket the terrain for miles. The large-scale remaking of the former dryland forest into rangeland for cattle was well underway by 1900, when ranchers on several of the islands began to broadcast grass seed from North America, Africa, and elsewhere. Now ecologists blame the grasses for preventing the regrowth of native tree seedlings. Most likely the original vegetation here was sparse under the trees, which were themselves spaced well apart as an adaptation to scarce groundwater. The introduced grasses inhibit recovery of the native forest in several ways. Grass roots spread widely and penetrate as deeply as several meters into the soil, competing aggressively with other plants for water and nutrients. Also, the roots of most grass species imported to

The last remnant of an ancient Hawaiian dry forest, a lone ʻ*ohe makai* tree *(Reynoldsia sandwicensis)* overlooks a nonnative grassland in North Kona on the Big Island.

Hawaiʻi release secretions that are toxic to other plants—a kind of natural chemical warfare waged in the soil (Smith 1985). Above the ground, the dense grass cover blocks the sunlight needed by native tree seedlings in order to grow rapidly whenever a soaking rain falls on this area—an infrequent occurrence.

Further threats include alien insects and mammals, especially rats and mice. Near the Big Island's Mamalahoa Highway, in the vicinity of Puʻuwaʻawaʻa, there is an old isolated *uhiuhi* tree growing on a steep, rocky embankment. Each summer, this tree produces a sizeable crop of oval pods with one or two healthy seeds inside. On several occasions we have observed that many of the seeds on the ground have been damaged and perhaps killed by the gnawing of rats or mice. It is difficult to tell if this seed predation took place entirely on the ground or if rats (presumably the arboreal species *Rattus rattus*) raided the treetop. This tree survived a recent fire that swept up the adjacent gulch from below, killing many large *kauila*. Fortunately the *uhiuhi* was not surrounded by thick dry grass, as were the *kauila* trees. But wild rodents are common in this area, and they are more difficult to control than fire or even the proliferation of the smothering alien grasses. Their destruction of the few seeds (and perhaps seedlings) that are produced by the dwindling numbers of species such as *uhiuhi* may be a major factor driving our rarest plants toward extinction.

The passage of the federal Endangered Species Act of 1973, followed by similar state legislation in Hawaiʻi, initially raised hopes for saving the remaining wild populations of rare plants. However, here as elsewhere, such laws offer only partial protection. While habitats in which rare species were living could no longer be routinely altered or developed and the plants themselves could not be deliberately destroyed, damaged, dug up, or otherwise knowingly disturbed, the law could not save any species from fire, aggressive alien grasses, or destructive animal pests. Without massive and expensive human help, most of Hawai'i's last patches of wild native landscape in the lowlands may well disappear in the next decade or two.

Preserving Hawai'i's Botanical Heritage

The task of saving Hawai'i's unique native flora involves two related but different goals: (1) to prevent the extinction of each of the many endemic and indigenous plant species in Hawai'i and, (2) to preserve or restore the natural native populations of these plants.

Saving Species

We believe that the best and most immediate way of saving Hawai'i's endangered plants is to cultivate as many native plants as possible. Already, a number of species that were near extinction such as *koki'o 'ula*, *ma'o hau hele*, *nānū*, and *'ōhai* are becoming common in urban settings because people decided they wanted to grow them in their yards, around their schools, and in their parks. While these species remain rare in the wild, cultivation has given them a new domestic home and an opportunity for future reclamation of their former habitats. If we are to prevent the extinction of other endangered Hawaiian plants, cultivation is a proven method. We therefore encourage all Hawai'i's gardeners to grow natives.

To increase the number of native species that we can successfully cultivate will require cooperation and the exchange of information among growers. This is why it is so important to establish a network with other people interested in native plants. One of the best ways of networking is to join a native plant group such as the Hawaiian Botanical Society, the Native Plant Society of Maui, and others (see Appendix 1). As a member of one of these groups, you can keep informed about new and better methods for cultivating native species (and perhaps share your own experiences). The groups also connect you with people who have surplus plants or seeds so that you can expand your native garden without having to obtain collecting permits or wait for infrequent native plant sales.

As with all enterprises, there are several ethics associated with the cultivation of native plants. The first of these (and, we would argue, the most important) is that native plants should not be removed from wild areas. The idea of taking a plant from its native habitat solely for personal enjoyment (or profit) should be unthinkable. But if the keeping of native plants becomes popular—as we hope it will—there will be the temptation for amateur gardeners and professional profiteers to dig up wild plants. If you find yourself tempted, remember these facts: (1) Taking native plants from wild areas without permission from the government (on state and federal lands) or the landowner (on private lands) is against the law. (Actually, the law prohibits not only taking plants but plant parts too, such as seeds, flowers, or cuttings.) There is a stiff penalty for violators. (2) Plants that are dug up from the wild rarely survive. It is virtually impossible to get their roots uninjured out of the typically rocky ground. Even if you do manage to get a significant portion of undamaged root, the shock of transport and transplanting is usually enough to kill the plant. (3) Taking wild plants deprives us all of the enjoyment of seeing our native flora in its unspoiled habitat. Nearly everyone enjoys seeing native plants in wild areas while hiking, exploring, or on a quiet walk along the seashore. Take the plants and you take away their enjoyment.

Should you collect seeds or cuttings? Many gardeners first experiment with plant propagation via cuttings. You see something you like in your neighbor's yard and you ask to take a piece of it. While this technique works well for many plants, we recommend that people interested in propagating native Hawaiian plants do so with seeds. By collecting only seeds from a parent plant, your impact on the plant or the population is almost always negligible (provided of course that you exercise moderation when collecting). Taking a cutting, any way you look at it, is an assault on the plant. By cutting the plant you leave it more vulnerable to diseases and pests. You also temporarily reduce its ability to manufacture its own food. This can stress the plant, again leaving it more vulnerable to diseases or pests. Because we still have much to learn about propagating native Hawaiian plants (their morphologies in early stages, growth rates, etc.), you can make valuable contributions to the science by starting with seeds and keeping good records of your progress. Sometimes, however, for species such as *koki'o 'ula, kulu'ī, naio,* or *'ōhi'a lehua,* propagation by cuttings is justified either because it is so much easier than seed propagation or because you want to maintain the genetic identity of the plant (e.g., orange flowers on a *koki'o 'ula*). Of course, remember to get permission whenever you are collecting.

Probably the biggest concern professional botanists and government agencies have about the cultivation of native Hawaiian plants by the general public is that people will start putting the cultivated plants back into wild places. While this initially sounds like a good idea, there are some very good reasons not to do so. (1) Cultivated plants frequently have unseen pests and diseases that are not present in the wild. By planting them out in the wild you may inadvertently spread these pests or diseases to the native plants you want to help. (2) Natural plant populations are normally composed of many genetically different individual plants. It is this diversity that helps them to adapt and persist when their environment changes. In contrast, cultivated plants are normally propagated from one or maybe a few original plants; their genetic diversity is small. When numerous cultivated plants are introduced into a wild population, their interbreeding can decrease the genetic diversity of the population. This decrease may harm the population by making it less able to adapt the next time there is a change in its environment. (3) Just as wild plants adapt to particular environmental conditions, cultivated plants change over time to better survive the conditions of cultivation. Cultivated plants are also selected by us for characteristics we find particularly attractive or useful. It is unlikely that a cultivated plant will have the same adaptations as a population of wild plants. So, when we introduce a cultivated plant into a wild population and genes are exchanged through crossbreeding, the wild population becomes less adapted and, therefore, less able to survive in its environment.

One of the current guidelines regarding the cultivation of natives, initiated by professional botanists and now strongly supported by state transport and commerce laws and regulations, is to cultivate only species or varieties endemic to the island you live on (i.e., gardeners on Maui should grow only native species that naturally occur on Maui; Oʻahu gardeners should grow only natives that naturally occur on Oʻahu). We have mixed feelings regarding this "rule." From a botanical and ecological point of view, the guideline is sound. Many native Hawaiian plants have genetically different populations living on the different islands (they often look different, too). While some of these differences may simply be the result of chance, others are almost certainly adaptations to the different environments in which each evolved. If we start cultivating all the varieties together on the same island, there is the danger that this natural variation will be lost and we will end up with, for example, only one type of *naio*—and it may not be well adapted to any natural environment. There is also the possibility that a disease or pest of one native plant may inadvertently stow away on a transported plant to another island, thereby exposing one island's natives to another island's pests. A third potential problem with cultivating natives from another island is that it will further obscure our understanding of how Hawaiʻi's many native plants evolved, particularly if people begin illegally planting these neighbor island natives in wild areas.

From a more pragmatic view, the guideline ignores the fact that over 80 percent of Hawaiʻi's residents live on Oʻahu. There simply are not enough people interested in growing native plants on Lānaʻi, for example, to ensure the survival of the many endangered and threatened plants found only there. We cannot move the people to the plants, so the plants must come to the people. Furthermore, cultivation is a means of preserving a species (or perhaps several varieties of a species), not wild populations (the second of the two goals mentioned above). Unless cultivated plants from other islands are indiscriminately mixed with natural populations, as in the case of illegal outplanting, they do not threaten the integrity of wild populations. Finally, it will be hard to persuade Hawaiʻi's public to follow a rule that nearly all the professionals break. For practical reasons, many botanists, land managers, and horticulturists employed by the federal, state, and city governments and by private organizations involved in conservation have had to import, propagate, and maintain natives from neighbor islands. Without their efforts, many more Hawaiian plants would now be extinct.

The Revised State Law Regarding Endangered Plants

There is considerable interest among Hawaiian gardeners, landscape architects, nursery workers, and others, in the revisions to the State of Hawai'i law regarding endangered plant species. We urge anyone planning to propagate, plant, or sell endangered species to contact the Department of Land and Natural Resources (DLNR) first so that you can avoid breaking the law. Following is an overview of the law and its implications.

- First, it continues to be illegal to collect plants, seeds, cuttings, and so on, from any noncultivated (i.e., wild) endangered plant on any land throughout Hawai'i (federal, state, private) without a permit from the State of Hawai'i DLNR.
- Second, in the past it was illegal to possess an endangered plant (or any part of an endangered plant) without a State of Hawai'i permit. This is no longer true. Now it is legal to possess an endangered plant (or parts) if the material is from a cultivated source. A cultivated source is a grower such as a state or city garden (e.g., Lyon Arboretum) or a private nursery or individual that has applied for and received a license from the State to give away or sell excess endangered plants. This grower is required to provide information to recipients about planting restrictions and to maintain certain records (contact the DLNR for the details). When you obtain an endangered plant from a licensed source, it will have attached a small numbered label (and usually an informational leaflet) that explains the planting restrictions. Most important among these restrictions is that the endangered plants you obtain from licensed growers—and anything you propagate using them—should not be planted out into wild areas where they could potentially threaten natural populations of the species with alien pests, hybridization, and so on. In other words, keep your endangered plants in your garden!
- Third (and this is the best news), if you want to give away seeds, cuttings, or plants you have collected or grown from your legally acquired plant, you can do so without getting a license or permit. If you want to sell this material, however, you must get permission from the State. This means that if your friend wants some seeds from your *ma'o hau hele,* our state flower, you can give them to him or her without fear of breaking the law. But please advise your friend, as you were advised, not to plant the plant in any wild areas.
- Fourth, a license from the State is still required to collect plants, seeds, cuttings, and so on, of nonendangered plants you encounter on any state land such as parks, forest reserves, or sanctuaries. You should contact the DLNR for the license.
- Fifth, federal laws regarding the collection, import, export, and so on, of endangered and threatened plant species remain unchanged. You should contact the local office of the U.S. Fish and Wildlife Service for further information on federal permits and restrictions.

Both coauthors have cultivated native plants endemic to islands other than our home island. We have done so to learn how they grow and because we saw their existence seriously threatened. We have always tried to be responsible in our activities. We urge you to make up your own mind about this guideline based on the arguments we and others will present you. Should you decide to grow natives from other islands, please always do so with knowledge and responsibility.

Preserving and Restoring Natural Populations

For many Hawaiian plants, the prognosis for unaided survival in the wild is not good. Indeed, chaos theory (sensitive dependence on initial conditions—see Lewin 1992) pessimistically (or perhaps realistically) would suggest that the changes to Hawai'i's ecosystems in the last several hundred years are so numerous and fundamental that those ecosystems cannot be restored to anything close to their former state. We can only hope that this is wrong. Ironically, for many years the prevailing attitude of state and federal agencies charged with protecting Hawai'i's native ecosystems, and especially the rare plants and animals within them, was somewhat hostile to public participation in saving Hawai'i's natural heritage. In fairness to these agencies, we must also point out that until recently the public had little interest in Hawai'i's native plants or their perilous state. Today, however, public interest is growing, due in large part to a renewed interest in Hawaiian cultural and natural history. These same agencies are now learning, as is the rest of the world, that there can be no lasting conservation of natural areas without public knowledge, concern, support, and involvement. Evidence of the State's change in attitude includes the recent revision of the State's endangered species law and its attempts to work with, rather than against, individuals who were previously considered ecological militants. But old habits are tough to break, and we believe that even more revisions in laws, regulations—and attitude—will be necessary before the public can feel a true sense of ownership in the lands it has entrusted to our state and federal governments.

Meanwhile, it is important to recognize the different roles that the general public and government agencies (or private conservation organizations) play in wildlife preservation and restoration. The agencies and organizations employ botanists and ecologists in conservation efforts, including native plant restoration. The agencies also enlist and coordinate outside professionals in such work. These professionals best understand the complex biology of Hawaiian ecosystems. They are aware of the problems and many of the solutions associated with preservation and restoration. Unfortunately, there are too few of them to do all the work that is needed. And, of course, their work perennially suffers from stringent and declining budgets.

In contrast, we the public may not know all the science, but we are many and we are ready to help. We can grow the numerous plants needed for a restoration project. We can get together on a Saturday morning and clear a site of introduced pest plants,

With one of the authors, volunteer students and faculty from Leeward Community College are restoring the native vegetation at a coastal site on Barbers Point, Oʻahu.

and we can plant a sizeable landscape in a single day. We can put up a fence to protect these plants from the goats, cattle, and pigs that might harm them. We can help the botanists and ecologists collect important data from the nursery or the field to better understand our Hawaiian biota. We can further educate ourselves about native plants and native ecosystems so that we can make some of the decisions about preservation and restoration. And finally, we can pressure the agencies, the organizations, and our legislators to let us do more to save our Hawaiʻi!

HOW TO GROW HAWAIIAN PLANTS

Handling Seeds

Many types of Hawaiian seeds store poorly (see Yoshinaga 1998). Therefore, for most Hawaiian natives it is best to start with freshly gathered fruits. The seeds from ripe fruits are best for sprouting. But in plants that naturally produce dry fruits or seed capsules, such as *kauila,* avoid mistaking year-old fruits that are still attached to the tree for the fresh crop. Whenever possible, collect fruits from more than one plant as the seeds of individual plants often vary in their viability. Examine the collected material for insect damage. Damage limited to the pulp or husk is no problem, but a seed with a hole bored into it—even a small hole—is nearly always dead.

Extract the seeds from ripe fruits soon after collecting and wash them in tap water to remove any attached pulp. Do not let the fruits rot because this often contaminates the seeds with destructive microbes. For short-term storage (one or two months), air-dry either the washed seeds or the intact fruits and keep them at room temperature or in the refrigerator. Drying the seeds or fruits will keep most spores of fungi and bacteria dormant. Do not store seeds or fruits in your freezer.

Sometimes, full-sized but green fruits yield viable seed (e.g., *'ala'a, 'ilie'e, kauila, pāpala kēpau*). Let these ripen over a period of one week to a month under clean, dry conditions. Protect the fruits from insects, and do not let them touch one another to prevent the transfer of disease or even undetected insects such as seed borers. Do not refrigerate the green fruits or seal them in a plastic bag; this almost always promotes the growth of microbes and causes rot. As they dry, the green fruits will shrivel and harden. This is normal, and you should wait until they are brown and dead in appearance before trying to harvest the seeds. You can then soak the dried fruits in water for a few hours. This will allow you to easily extract the seeds from the softened pulp.

For a few plants, green but full-sized fruits (or fruiting stalks with full-sized, green fruits) can be ripened by keeping the stems of the fruit or stalk in water for a week or more. This works well, for example, with the long-stemmed cucumber-like fruits of *maiapilo (Capparis sandwichiana)* and the fleshy stalk of *hala pepe (Pleomele* sp.*)* with its many cherrylike fruits (see discussions of these plants in the next section).

Scarifying a seed is the process of softening or partially breaking down the seed

The yellowing skin and soft, red pulp of this ripe *maiapilo (Capparis sandwichiana)* fruit tells you that the many dark seeds within are ready to be harvested, cleaned, and sown.

coat. This often speeds germination. The reason is that if a seed is unable to take up water from its environment, it will remain dormant. For many seeds—especially those of dry-adapted species—there is a thick and waterproof seed coat surrounding the embryo. Chemical inhibitors to sprouting may also be concentrated in the seed coat under the waterproof coating, thus preventing it from leaching away quickly. In nature, it often takes years before these seeds sprout, although the process can be hastened by an exceptionally rainy season or because of natural scarring by rocks in the soil. By scarifying a seed, you circumvent the slow natural breakdown of the seed coat, the barrier to sprouting.

Usually, vigorously scratching the seed's surface in one or two places with a penknife blade or with medium sandpaper is all that is needed to break through the waterproof barrier. Avoid damaging the embryo with excessive mechanical scarification. Hot water scarification, convenient with small seeds, involves placing the seeds in hot water and letting the water and seeds cool to room temperature. Opinions vary on the optimum water temperature (anywhere from 46°C to 82°C, 115°F to 180°F); we usually use water at about 66°C (150°F). Not all seeds sprout more quickly when scarified. For example, seeds from *lama, maiapilo,* and *nānū* germinate just as quickly without scarifying. When in doubt, try experimenting by scarifying half the seeds and comparing their germination with nonscarified seeds.

When planting seeds, it helps to recall something about the conditions of nature in Hawai'i today. The influx of plant-destroying insects, fungi, bacteria, and viruses has accelerated dramatically since the time of Captain Cook. These destructive alien organisms have especially invaded the soils of Hawai'i's lowlands. Because Hawaiian plants are most susceptible to these threats when they are very small, you should not even think about using soil from your yard to sprout native Hawaiian seeds or grow native seedlings. Instead, always start your seeds in artificial soil materials.

The following is our standard germination technique. It works well with many Hawaiian species, including those with seeds and seedlings susceptible to microbial diseases.

1. Extract the seeds from ripe fruits within one or two days after collection (or later from the full-sized green fruits of some species; see above and next section for detailed discussions of these species).
2. Immerse the seeds, at room temperature, in a solution of one part household bleach to nine parts water for approximately half an hour. With small seeds less than 3 mm (⅛ in) in diameter, dilute the bleach by about half. This step should be used with caution because some seeds (for example, *hao*) are damaged by the bleach. Use it when you suspect the seeds may be contaminated by microbes, such as seeds coming from dry fruits or fruits collected from the ground. Conversely, soaking seeds in a dilute bleach solution stimulates germination in some nonnative seeds such as rice (Mikkelson and Sinah 1961) and oats (Hsiao and Quick 1984).
3. Remove the seeds from the bleach using a colander or a piece of clean window screen. Otherwise use clean forceps (or tweezers). After pouring off the bleach solution, use the same container to continue soaking the seeds in tap water. Fill the container with enough clean water to just cover the seeds (this allows plenty of oxygen to diffuse into the water). Soak the seeds for about a day. Covering the container loosely with clean plastic wrap is a good precaution at this stage.
4. Prepare a seedbed by placing a layer 3–6 cm (1–2 in) thick of new vermiculite (available from garden shops) in a clean container with a well-perforated bottom. (It is good practice to have earlier submerged the empty seedbed container for an hour or so in a bucket of bleach solution; make this a little stronger than what you would use to wash your sheets or towels.) Wet the vermiculite until water drains out the holes in the bottom of the container. For a convenient container, try plastic tofu tubs. The tub's bottom is easily perforated with the blade of a small knife. Make many small holes to guarantee that the seedbed has good drainage. If you do not like tofu, cherry tomato containers also work well.
5. Place or scatter the seeds on top of the vermiculite. Use clean forceps or very clean fingers to minimize contaminating the seeds with microbes. With seeds larger than a few millimeters, such as those of *'iliahi* or *kōlea,* it is a good idea to embed them in the vermiculite.

6. Obtain a handful of fresh green moss (this is the noncrumbling, greener, more alive-looking version of peat moss). It can be purchased along with vermiculite at nearly any garden shop. Although this material is not sterile, it is an excellent covering for the seedbed. It will hold a lot of water, keeping the seeds moist, and it also permits air to circulate around the seeds. The moss itself contains antifungal substances that reduce the incidence of damping-off in seedlings (Fleming and Hess 1965). From our own observations, the moss also seems to reduce the incidence of mold on seeds. Holding the moss in your (clean) hands, soak it thoroughly with water and shape it into a mat about 2–3 cm (1 in) thick that closely fits into the top of your seedbed container. There should be room in the seedbed to gently press the cake of soggy moss down to contact and cover the vermiculite and seeds.
7. Finally, water the seedbed, with its moss cover, once more until water drains out the bottom of the container. From then on, water the seedbed about every two days or whenever the top of the moss dries out.

Seedlings

Check under the green moss every few days for sprouting seeds. Over time, the moss tends to consolidate into a fairly coherent pad that can be peeled up en masse and replaced after examining your seeds. Look out for seeds that adhere to the moss; however, most will stay put on the surface of the vermiculite. By consistently checking the seeds every few days you can find and remove dead seeds early. Despite the moss' antimicrobial properties, dead seeds often become coated with irregular, white, sometimes fuzzy-looking growths. Occasionally this also happens to healthy seeds—usually because of bits of pulp left adhering to the seeds—but in such cases, the bacterial or fungal coatings are typically sparse and fade away after a week or two. If you are in doubt about the health of a seed, test it by squeezing it between your fingers. Rotting seeds usually crumple readily, while healthy seeds do not. Be sure to wash your fingers or forceps after they touch rotting seeds to avoid contaminating healthy seeds or sprouts with harmful microbes. You can rub off superficial mold on healthy seeds with your fingers under a dilute bleach solution or warm water. Sometimes making a new moss cover or exposing the seeds on the seedbed (with the moss mat removed) to sunlight for a minute or two also helps to eliminate persistent mold. Usually, however, if you start with clean, healthy seeds you will find that they remain mold free under the moss cover.

Given the opportunity, many types of insects, insect larvae, and other pests will attack seeds and sprouting seedlings. Indoors, fungus gnats (Mycetophilidae and Sciaridae) and cockroaches are common culprits. Outside, the risks are even greater, owing in part to highly specialized seed predators (see next section on *hala pepe* for an example). Covering your seedbeds and young plants with screens provides some protection, but not from everything. For example, a tiny borer that slips through window screen can kill *lama* seedlings. Birds, mice, toads (Bufos love to dig into and hunker down in moist vermiculite), and curious cats will also disturb or destroy seeds and seedlings in accessible seedbeds. Adapting to your own particular surroundings and calling upon your own ingenuity are the keys to foiling such pests.

The first sign of sprouting is normally the emergence of the root. Initial root growth can be slow in some plants, such as *kōlea*, or quite fast, as in *kauila* and *lama*. Roots that emerge from seeds on the surface of the vermiculite soon turn downward. This is the time to begin thinning the moss layer, making it easier for the new shoot to grow upward. Frequently, once the first few seeds sprout, many others follow within several days. In some species such as *maile* and *māmane*, the shoot and root appear almost simultaneously and develop at comparable rates. In other Hawaiian plants such as *kōlea*, *kōpiko*, and *lama*, the shoot takes much longer to develop, or at least to free itself from the rest of the seed. For most species, you should let the seedlings develop in the seedbed for three to four weeks after sprouting. During this period, remove most but not all of the moss from the seedbed; seedlings seem to benefit from a little moss surrounding their bases. Sprouts from large seeds such as those of *ʻalaʻa*, *ʻaulu*, and *uhiuhi*, obviously, must be removed early from the small seedbed before their roots become cramped and intertwined. Others such as the two *kauila* species and *lama* also seem to do better if they are removed from the seedbed and carefully potted as soon as they have 1–2 cm (less than 1 in) of root.

During the first month or so after sprouting, proper watering is very important—but do not fertilize. The young Hawaiian seedlings use their stored nutrients during this period and in many cases adding fertilizer does them more harm than good.

Propagation by Stem-Tip Cuttings

It has been our experience that, unlike stem-tip cuttings from many wet forest or even mesic forest plants that consistently root using a frequent misting system, stem-tip cuttings from dryland and coastal species often rot and die using this technique. Instead, dryland and coastal species seem to root best when placed in a humid environment with infrequent misting. We have used the following technique repeatedly with *ʻākia*, *hinahina*, *ʻilieʻe*, *kokiʻo ʻula*, *kuluʻī*, *naupaka*, *nehe*, *naio*, and *pāʻuohiʻiaka* with very good success.

Use freshly cut 10–15 cm (4–6 in) stem tips. If you need to transport the cuttings, keep them cool and in a closed container such as a plastic bag with a small amount of water. Wash the cuttings well under tap water to remove any insect or fungal pests. Remove the leaves from the lower half of their stems. Insert each stem in its own pot, filled with a 1:1 mix of moist vermiculite and perlite. Treating with rooting hormones is not necessary for any of the above species. We also recommend against removing or trimming any of the remaining leaves to reduce transpirational water loss (see why in the next section on *kokiʻo ʻula*). Place the cuttings in a small covered aquarium or other clear container. Most recently, we have started using 32 oz

Stem-tip cuttings from dryland plants tend to rot under a frequent misting system. We have had much better results placing the potted cutting in a moist chamber, such as this plastic cup, and lightly misting every one to three days.

clear plastic cups with either dome or flat lids for each cutting. Place the containers with cuttings under artificial or indirect light and, with a spray bottle, lightly mist the cuttings once a day. We discourage placing more than one cutting in a pot as fungus from one dead stem tip can infect and kill the surrounding cuttings.

Watch the cuttings closely and destroy any insects that appear within the container. Also promptly remove any dead leaves before they begin to rot or become moldy. Some of the above species, such as *hinahina* and *nehe,* will begin rooting in as little as one week. Others will take longer; three or four weeks for *koki'o 'ula* and *naio.* If after three months the cutting still has not rooted, chances are it is not going to and you should probably start again with fresh material.

Potting

Most often, for the first transplanting of seedlings out of the vermiculite seedbeds, we use either a two- or three-part potting mixture. The two-part mix is made of equal volumes of vermiculite and small- or medium-graded black cinder (this is the lightweight, porous rock sold by nurseries or garden suppliers, not the hard, dense gravel used in concrete and walkways). The three-part mix is made of equal volumes of vermiculite, perlite, and peat moss. Both potting mixtures have excellent drainage, yet the vermiculite and peat moss also hold water well. Unopened bags of vermiculite and peat moss are reputed to be clean and free of plant disease agents; the former may be sterile owing to the high temperatures reached in its manufacture. We sometimes add an additional ingredient to the mix: powdered sulfur (available from most garden suppliers). It seems to protect a wide variety of young Hawaiian plants against many diseases. Add about half a teaspoon (2.5 mL) of sulfur powder per gallon of dry potting mixture and mix it in thoroughly. (Note: Some people are allergic to sulfur; consider wearing gloves when handling the sulfur or any potting mix containing it.)

Scoop the dry potting mixture into clean pots of proper size for your seedlings (5–8 cm, 2–3 in diameter for most), and wet the mix thoroughly. One technique we recommend is submerging the pots in water about two-thirds their depth. About a minute is usually sufficient. Besides wetting the soil, this method seems to help the sulfur bind to the potting mix in the upper half of the pot. Over the next few weeks, it will filter down as you water the plants.

Remove a seedling from the seedbed by inserting a small knife blade under the vermiculite and lifting as you gently pull up on the plant's stem. The root will come

Sowing in clean, new vermiculite helped prevent bacteria and fungi from attacking these germinating *hala pepe (Pleomele hawaiiensis)* seeds.

Newly sprouted seedlings of *maiapilo (Capparis sandwichiana)* emerging from a seedbed of vermiculite and green moss.

up nearly intact, usually with soft, moist flakes of vermiculite clinging to it. Make a hole in the moist potting mixture and lower the root into it; then carefully fill in around the root as you water from the top with a small stream of water. Gently press down on the soil around the rim of the pot, and the job is done. Do not worry if the mix looks too chunky or rocky with cinders; as long as you do not let the mixture dry out, your plants will thrive.

Good husbandry following potting involves keeping the plants well watered and, for three or four days, partly shaded. Even dry-adapted species that thrive in full sunlight should not be left out in the noonday sun after being transplanted. Actually, it is a good idea to do your transplanting work late in the afternoon, leaving a stress-free period for the plants to recover overnight. Do not fertilize the newly transplanted seedlings for perhaps two or three weeks. New growth will appear as the plant adjusts to its new soil. When it does, begin a regime of light fertilization to hasten the seedling's growth through this vulnerable period.

When you water, continue until the water flows freely out of the bottom of the pot. When the plants are small, most of the water in the pot slowly evaporates from the potting mix and you may have to water only every three or four days. As the plants grow, however, they will begin drinking more and you will need to water more frequently. A large plant in a small pot can kill itself by dehydrating the soil in just a few hours on a hot summer day. With the potting mixtures described above, it is probably impossible to overwater the plants described in this book. If you are unsure about the optimum delivery of water, however, you can hedge your bets by tilting the pot when watering so that the water flows through on only one side. This will set up a moisture gradient in the lower reaches of the pot, allowing the plant's roots to seek their "comfort zone." (Several Hawaiian dryland species such as *'aiea, hala pepe, maua,* and *uhiuhi* thrive under this watering practice.)

One problem with frequent watering is that moisture-dependent diseases can attack vulnerable young plants, causing their stems to rot near the soil line. Root-destroying fungi are also more likely to appear in wet soil. Since moisture promotes these diseases, such as damping off, one solution is to keep the surface soil dry. Water the plant by partially immersing the pot for a minute or two in a container of water. This works well for *kauila* seedlings.

Fertilize young native plants with moderation. It is better to underfertilize than to use too much because excess nutrients can cause water stress and wilting (potentially killing new seedlings) and encourage damping-off. Some species such as *ʻākia, hala pepe,* and *ʻaiea,* as seedlings, grow well with nothing but small nitrogen additions to the potting mixture. Breakdown of the volcanic cinders seems to provide adequate nutrients in other categories. Most species respond best to liquid fertilizers such as Miracle-Gro (diluted to about half the recommended strength) or controlled release fertilizers such as Gaviota (14-14-14). Use the diluted liquid fertilizer once or twice a month by saturating the potting mix until you see runoff from the bottom of the pot. For the controlled release fertilizer, scatter a few pellets on the surface every two or three months.

A number of Hawaiian plants in pots, including some that naturally occur in hot, dry regions, are sensitive to high soil temperatures. This problem is particularly evident in the summer when air temperature and thus the soil in outdoor pots, unlike that even a few centimeters below the surface of the ground, reaches temperatures exceeding 30°C (86°F) for up to several hours during the day. These plants, even when well watered, will enter a summer stasis and stop growing for as long as several months. Fertilizer will not stimulate growth at this time and may even be harmful, causing dehydration. If you see this effect, measure the soil temperature during midday to confirm its high reading. Then try moving your plants to a cooler location that still provides adequate sunlight.

Planting Out

This is often the most difficult phase in cultivating Hawaiian native plants. A plant raised indoors or in otherwise sheltered conditions may have a particularly difficult time in making the transition to a permanent site in the landscape. Problems as simple as stem strength vis-à-vis resistance to wind can loom large at this time. It is a good idea to provide the plant with some time to acclimate to its new surroundings. Do this by keeping it in its pot for a few weeks after moving it into the open air. Later, move the plant progressively into more exposed areas representative of the site where you ultimately will be planting it out. Watch for the arrival of insect pests and practice remedies before you commit your hard-raised native to the ground.

Perhaps the most valuable rule to learn when planting out your natives is to place the plant in the appropriate habitat. *Hāpuʻu* ferns will not grow well (emphasis on the word *well*) on a hot leeward beach no matter what you try. You can modify the habitat by increased watering, shading, or soil augmentation, but this only adds another level of maintenance that may or may not lead to success. Most of the species detailed in this book are the dryland or coastal species that probably occupied the habitat that humans now monopolize—the dry leeward side of the main islands. While we have not done extensive testing, we believe that if you plant a dryland forest tree in a dryland area—lower Makiki for example—it should do well. In contrast, if you try to plant the same tree down near Ala Moana Beach, it probably will not thrive. Of course, future researchers and gardeners will discover exceptions, but it is always best to look at where you live and ask the question, "Could this species once have inhabited this area?"

When you are finally ready to transfer your carefully nurtured plants from their pots to their ultimate homes in the landscape, we suggest several important principles and practices to follow.

Hole size is important. Dig a large hole, even for a small plant. For a tree species or a large shrub, try a hole a meter (3 ft) in diameter and half a meter deep at its center. By digging a large hole, you do several things. First, if the original ground was very hard, you loosen the soil. The new plant will have an easier time spreading its roots through the loose soil. Second, roots need oxygen (for aerobic metabolism) to grow. Digging a large hole increases the amount of aerated soil surrounding the new plant. Third, a large hole has more inter-soil spaces where water can be retained. And fourth, a large hole eliminates, at least temporarily, competition from surrounding weeds and underground roots. Experiments conducted by coauthor Koebele comparing hole size clearly showed that plants in large holes (1 m diameter, ½ m deep) were much more drought resistant than plants placed in smaller holes (½ m diameter, ½ m deep) for up to a year.

Soil augmentation is of questionable value. A number of Hawaiian horticulturists (Nagata 1992, National Tropical Botanical Garden 1992, Bornhorst and Rauch 1994, Bornhorst 1996) have suggested augmenting the soil in the planting hole if the native soil has poor drainage. Suggested drainage materials have included cinder, perlite, or compost. The suggestion seems to have originated from experience with potted native plants, which indeed do better in high-drainage potting mixtures. However, in experiments conducted by Koebele we could find no advantage to soil augmentation. In fact, several plants, after having 50 percent of the soil replaced with cinder, actually grew more slowly than control plants. A recent article in *Hawai'i Horticulture* (by C. A. S. Bebb, 1998) also recommends against amending the soil in a planting site for most situations. While we do believe that good drainage is important for many native dryland species, we do not believe soil augmentation is the answer. Instead, we suggest that if the proposed site for planting out is an area where water tends to sit after rains, you should either find another location or build up the site with surrounding native soil to eliminate the standing water. Augmenting a poorly draining hole with highly porous material such as cinder tends to create a "bucket" of standing water surrounding the roots of your dryland plant. This is just the opposite of what you want.

On the other hand, there is merit in augmenting your soil if you suspect the soil chemistry is incompatible with the species you wish to grow. Coauthor Culliney has been able to dramatically improve the growth of some dryland species by augmenting the sandy, calcareous soil in his backyard with a more volcanic mix collected from a nearby area. This might be further evidence (by way of successful circumvention) for our first rule of planting out: "Place the plant in the appropriate habitat."

Mulching can be important. The use of mulches appears beneficial to many Hawaiian trees and shrubs, improving conditions in three ways. First, mulch can greatly reduce the frequency of watering for new plants. This is because the mulch provides a highly effective barrier to evaporation from the soil. Second, mulch can significantly reduce the encroachment of weeds and grass into the immediate area surrounding the native plant, thus reducing their competition for water, sunlight, and so on. Third, mulch helps prevent hardening of the surface soil, allowing rain and air to more effectively enter the soil surrounding your plant. Unfortunately, one disadvantage of mulch is that it sometimes encourages the buildup of large populations of wood roaches, sowbugs, millipedes, slugs, and other pests. Young plants with soft stems such as *hala pepe* are extremely vulnerable to chewing damage from these animals. If they are common in the area, try keeping the mulch 10–20 cm (4–8 in) away from the base of the plant. You may also have to use an insecticide on the ground immediately around the plants. Diazinon granules effectively eliminate most of these pests.

Watering is essential. New plantings are highly susceptible to desiccation. Water the planting hole thoroughly before and immediately after planting. Follow this with a daily examination of the plant for any signs of drought stress and water if necessary. After about two weeks, you can become less vigilant with your examinations; examine once or twice a week. For many dryland species, supplementary watering becomes unnecessary after about a year. Of course, this will depend on the particular species you plant, and you should refer to the more detailed watering notes in the remainder of this book.

Fertilization is important. Nearly all the natives we have planted out respond well to fertilizer. We recommend general all-purpose types with either equal amounts of nitrogen, potassium, and phosphorous or types with increased phosphorous because many Hawaiian soils are deficit in this element. Follow the directions that come with the fertilizer. The best time to fertilize is during the wetter months (October to May); during the summer, many dryland species such as *'ohe makai* and *wiliwili* slow their growth or even lose their leaves. Some Hawaiian naturalists (John Obata, per. comm.) have suggested that many of Hawai'i's plants probably evolved under conditions of low-level, continuous, or frequent fertilization from native bird droppings.

Temporary shading can help. Exposure to direct sunlight during the first few days after planting out can stress even dry-adapted plants. Providing them with a little shade for a week or so will help them to become established with minimal wilting. In some cases, shading may be necessary only during midday when heat and water loss are most severe.

Pests and Diseases

The problem of pests attacking native plants can be particularly severe in Hawai'i because so many species of plant-devouring insects have colonized the Islands from all over the world. This onslaught of alien species has steadily intensified since the arrival of Captain Cook in 1778. Ongoing studies show that, despite growing awareness and attempts to control the problem, more than twenty new alien insects become established yearly (Beardsley 1979; Funasaki et al. 1988). Many of the introductions are first detected around Hawai'i's military air bases, despite Defense Department regulations that mandate inspection and control measures. (Of the several known incidents of brown tree snakes carried to Hawai'i from Guam, all have been via military aircraft. How much easier it is to inadvertently transport alien insects!) These invasions are often silent, with little attention from the media—unless an alien stowaway is already notorious. Once the recently arrived insects are detected they have usually built up breeding populations that are impossible to eradicate. Also, until very recently, a lack of strict control over the movement of agricultural products within the state of Hawai'i has allowed the spread of many plant pests and diseases from island to island.

Of course, it is primarily the lowlands of the main Hawaiian Islands that have been invaded, largely by tropical pest species from elsewhere in the world. Ironically, during the last two hundred years of agricultural development and ecological change in Hawai'i, many native insects have become extinct. As a result, in a few upland regions, away from long-term agricultural zones or older residential developments where crops and gardens have served as incubators for myriad alien pests, native plants may actually have fewer pests than in prehistoric times. But losses of native insects have also included mutualistic (or facilitator) species, so it cannot be said that, even in remote habitats, native vegetation has made any gains.

Attacks on plants by insects, snails, and other small animals fall into several categories. The most obvious is leaf-tissue consumption. Extremes involve species such as the Chinese rose beetle *(Adoretus sinicus)* that can develop dense populations and may virtually defoliate a shrub overnight. Some plants seem to cause pests to behave in unexpected ways. The large alien slug *(Vaginulus plebeius)* usually feeds nocturnally at ground level on decomposing leaf litter as well as green foliage, but this pest also routinely climbs up a meter or more to devour the leaves of Hawaiian natives such as *koki'o ke'oke'o* and *'a'ali'i*.

(Note: A very useful source of information on plant pests and diseases in Hawai'i is the University of Hawai'i's College of Tropical Agriculture and Human Resources Integrated Pest Management Program. You can contact them directly at the university or visit their web site, currently at: http://www.extento.hawaii.edu.)

The Worst Enemies of Hawaiian Native Plants

Foliage-Consuming Pests

While these pests cause the most obvious damage to plants, they are not always the most menacing in native gardens and landscapes. Foliage eaters sometimes render large shrubs and trees unsightly (usually temporarily), but they rarely attack with life-threatening severity. Young plants, however, can be completely denuded and killed by some foliage-eating pests. Below, we describe some of the common and more threatening defoliators that may be attracted to your Hawaiian plantings.

Chinese Rose Beetle *(Adoretus sinicus)*: In many coastal and lowland areas, both leeward and windward, the Chinese rose beetle can be a serious threat to a variety of native plants. The females emerge from the soil at night, often after a heavy rain, and devour the leaves of nonresistant species such as *āulu, koa, koki'o 'ula, ma'o hau hele, 'ōhai,* and *wiliwili* (see photo). Males, which are much less destructive, locate the females (probably following a mating pheromone released by the female) and mate with them on the plant. Eggs are laid in the soil and develop into white grubs that do not appear to harm native plants. The grubs feed on decaying plant material (Williams 1931). The entire life cycle from egg to adult takes five to eight weeks.

The most effective defense against the Chinese rose beetle is to plant the susceptible native under a street, yard, or porch light. The effectiveness of a night light is truly remarkable. We have planted sibling *koa, ma'o hau hele,* and *āulu* so that one plant was under a night light and the other was not. While the leaves of the plants in the darker site were mercilessly chewed to the petiole (leaf stem), the plants under the light remained virtually untouched. Unfortunately, it is not always possible to light up your backyard just to protect your native trees and shrubs. And such a deterrent would obviously be impossible for native wildland restoration. The second most effective deterrent we have discovered is a systemic insecticide such as acephate (e.g., Orthene or Isotox). Under dry conditions, a single spraying of acephate can remain effective for a month or slightly more depending on the plant species. But frequent rains can quickly diminish the acephate's effectiveness, and you may have to spray as often as every week.

Many Hawai'i gardeners have reported a cyclic pattern of Chinese rose beetle damage to their plants. This is probably related to the insect's life cycle and ecology. We have noted that the most severe damage usually occurs during the fall and spring. We have also noted that a heavy rain often precedes a particularly destructive evening, and that moonless nights are the worst. By keeping notes on severe damage

Pests and Diseases

Chinese rose beetles *(Adoretus sinicus)* go to work destroying the leaves (phyllodes) of a *koa*. The beetles live in the soil during the day, coming out at night to devour the leaves of susceptible native plants.

periods, temperature, and rains, it is possible to time chemical intervention and thereby minimize insecticide use and beetle damage.

Finally, tall grass surrounding a susceptible native plant may also deter Chinese rose beetles. We have seen this effect numerous times in a small grove of planted *wiliwili* trees. In the tall grass, we see little damage to the trees' leaves. Whenever the surrounding grass is cut, however, there is a dramatic increase in leaf damage from Chinese rose beetles. Perhaps the tall grass acts as a physical barrier to the flying females in search of a meal. In any case, our observation may be useful for those involved in the restoration of dryland forests and shrubland. (We doubt that most Hawai'i gardeners will be willing to let their backyard grass grow waist high just to protect their Hawaiian natives from beetles.)

Small Gray Weevil *(Myllocerus* sp.): A second serious foliage-eating pest is a small gray weevil. This insect is 3–6 mm (¼ in) long and seems to appear from nowhere on the young leaves of several susceptible Hawaiian natives, including *Achyranthes, koa, āulu, 'ōhai, kauila,* and *nanea*. They multiply rapidly and while individually they do not cause much damage, collectively they can devour all the young leaves on a tree or shrub. Fortunately, the weevil does not appear to feed on the apical meristem (the growing stem bud) nor on older, more lignified leaves. Controlling the weevil is not difficult; two to three sprayings with insecticidal soap or acephate will eliminate them—temporarily. Unfortunately, for several Hawaiian plants, repeated infestations are the norm.

Other Leaf Eaters: A wide range of other leaf-feeding insects appear from time to time on native plants. Often, their attacks are sporadic and short lived. For example, flea beetles (Subfamily Alticinae) leave many kinds of plants looking as if their leaves had been shot through by sand grains. Fortunately, the damage by these tiny insects is rarely severe or long lasting. Some other insects are specialists, even on Hawaiian plants. For example, larvae of the cabbage butterfly *(Artageia rapae)* attack *maiapilo*. The green "cabbage worms," if not noticed in time, can quickly defoliate young plants.

Grasshoppers, both native and introduced species, are attracted to and feed on some native plants. During an exceptionally dry period in the winter of 1998, these insects were responsible for defoliating several young *wiliwili* trees planted out near Hanauma Bay, Oʻahu. Grasshoppers at this same location were also responsible for some moderate foliage damage on *naio*.

Sucking Pests

The extent of damage from these pests depends largely on their numbers. In large numbers, they can threaten the life of a susceptible plant. For example, large numbers of scale insects on young *alaheʻe* or *ʻaʻaliʻi* or large numbers of mealybugs on *ʻōhai* can kill the plant. In contrast, a few scale insects on more resistant species such as *Achyranthes splendens* often disappear over time, probably due to natural predators. Another cause of damage by some types of sap-sucking insects is that their saliva contains substances that are toxic to plants. Necrosis, or tissue death, is the result after these kinds of pests feed on fluids from leaves or stems.

Scale Insects, Mealybugs, and Aphids: It has been our experience that the most serious threat from these kinds of sucking insects is when ants introduce them to a plant and protect them from predators. Usually you will notice the ants before you detect the sucking insects they are tending (the ants are attracted to sugary secretions produced by these kinds of sap suckers). The cure is to either eliminate the pest (the ants leave when the pest is destroyed), control the ants (and allow natural predators to destroy the pest), or eliminate both. The pests vary considerably in their resistance to treatment. You can quickly kill most aphids on foliage by spraying with a dilute soap solution. Mealybugs and especially scale insects such as the common yellow-green scale *(Coccus viridis* sp.*)* are tougher (sometimes protected by a waxy coating) and a single spraying may not eliminate them. Very dense infestations of sucking pests on plants such as *ʻōhai* with its complex leaf structure may require repeated treatment with insecticidal soap or acephate. The spraying also kills any ants present.

When your plants are small, controlling ants is fairly easy with a nontoxic, sticky

Scale insects can sometimes be difficult to get rid of, particularly if the pests are protected and farmed by ants.

resin called Tanglefoot or Tangletrap, obtainable at garden shops. Just encircle the plant's stem, well above the ground, with a small ring of the sticky stuff. This prevents the ants from reaching the foliage and defending their "herds" of sap-sucking insects against natural enemies such as ladybugs. Without protection, the suckers soon disappear. You can also protect larger plants using this technique, until they begin to grow up against other plants, a fence, or any object that permits the ants to reach the foliage by multiple routes. Eliminating ants can be more difficult than dealing with the associated pests. Depending on the species of ant, you may have to use different treatments; hydramethyinon (AMDRO) for fire ants, Diazinon for other species of ants.

In extreme cases (protecting your last specimen of a rare species?) it is easier to repeatedly destroy the pest than to eliminate the ants. This often requires two to four weeks of vigilant treatment to make certain you kill all the pests, including those reintroduced by the ants.

Perhaps the most insidious infestations by sucking pests are those on roots. Certain aphids and mealybugs specialize in sucking root fluids and build up large populations below ground, typically in association with ants. Potted plants, as well as those in the ground, may be subject to this kind of sneak attack. Symptoms are slowing or cessation of growth, wilting, and yellowing of foliage—if these changes cannot be attributed to other causes. A close look around the base of the stem will often reveal ant activity, and, if you dig carefully with a small blade, you may find the sucking pests themselves. The culprits are often just a centimeter or two below the surface; they are generally white or yellow and clinging to the plant's roots. The most effective treatment seems to be scattering Diazinon granules around the plant's stem;

if you can eliminate the ants, your plant may have a chance. Otherwise, a severe infestation will often kill a young plant.

Stinkbugs: These are hard-shelled insects that resemble beetles; they possess glands that release an acrid-smelling fluid when they are physically disturbed. They may appear nearly any time of year and are attracted to a variety of Hawaiian plants, where they suck the juices from young stems and petioles. They deposit their eggs on host plants; the crawling young bugs develop into the winged adults, and the life cycle may go through multiple generations on a single plant.

There are at least two species of stinkbugs that pose a threat to native plants. The green stinkbug *(Nezara viridula),* as its name implies, is bright green and about a centimeter (⅜ in) in length as an adult. Immature bugs are green as well but also have a chainlike pattern of reddish brown spots on their upper surfaces. On our plants, we have seen this stinkbug only in low numbers and, therefore, it has not been a serious pest; we simply kill the bug by hand when we encounter it. In contrast, the smaller black stinkbug *(Coptosoma xanthogramma)* can be a major threat to Hawaiian species in the bean family, particularly *wiliwili,* *'ōhai,* and *'āwikiwiki.* As an adult, this bug is smaller than its green cousin and sometimes mistaken for the beneficial black lady beetle that also commonly lives on the same plants. (Lady beetles eat aphids and other pests and you should leave them on your plants.) Young black stinkbugs are flightless and not black but bright green. You normally see them at the bases of leaves, feeding on the plant's sap. The black stinkbug has a more flattened body than a lady beetle and its coloration is slightly different, but if you are not sure, just pick up the bug and sniff. There is no mistaking the sharp, noxious odor of this bug that deserves its common name.

Black stinkbugs can reproduce quickly, establishing a large population in just a few weeks. Their feeding causes scarring on the plant's stems and near the bases of leaves. Larger plants, even 2- to 3-meter-high *wiliwili* trees, may be so stressed that they drop their leaves prematurely. Smaller shrubs such as *'ōhai* may be stunted or killed. Because insecticidal soap has little or no effect on black stinkbugs, you should spray with a more toxic insecticide such as acephate or malathion to reduce the population. Even then you may need to spray more than once to eliminate this pest. On small plants, we recommend inspecting the stems carefully and removing the bugs by hand.

Leafhoppers and Treehoppers: Leafhoppers are tiny, spindle-shaped insects that infest the leaves of many kinds of native and introduced plants in Hawai'i. Typically, these pests prefer the undersides of leaves, where they pierce the surface and feed on the plant's tissue fluids. A new infestation may escape notice until the foliage is disturbed. Then, even a light brushing will trigger a spattering or, worse, a cloud of the 2–3 mm (⅛ in) adults to erupt briefly into the air. The little bugs immediately land again nearby and seek new hiding places. Carefully turning over leaves will reveal the flightless young leafhoppers; they typically react to exposure with a rapid sideways movement.

Perhaps the most notorious of the leafhoppers is the two-spotted leafhopper *(Sophonia rufofascia),* which has an extremely broad host range. However, a related green leafhopper *(Empoasca* sp.*)* is found on many of the same plants. Leafhoppers produce plant toxins in their saliva. Affected plants exhibit a condition called "hopper burn"—usually a spotted appearance on the leaves, but in severe cases there is a general yellowing of the foliage. Severe hopper burn sometimes kills whole branches and perhaps whole plants. Protecting your plants from leafhoppers can be difficult. While spraying insecticidal soap on specific plants may nearly eliminate one infestation, there is a good chance these bugs will reappear quickly from other host plants nearby. They are especially hard to get rid of on ground-hugging plants such as *nanea,* a common host. Fortunately, leafhoppers may nearly disappear for up to months at a time, giving hard-pressed plants time to restore their vitality. Also, in our experience, many lowland Hawaiian plants are only lightly damaged by leafhoppers.

Treehoppers, such as the common *Vanduzeea segmentata,* live on stems. They are larger than leafhoppers and shaped like pointed hats. The head is large and the abdomen tapers to a blunt point. These pests often gather in small clusters around leaf petioles or on young stems, which may sustain severe scarring. They are also commonly tended by ants. These insects, too, seem to cause leaves to yellow and drop off. They can be a serious threat to young *uhiuhi,* and they also attack *maʻo,* stunting its growth. Successful treatment, however, is fairly easy. Control the ants with a ring of sticky resin around the plant's stem and the treehoppers will often disappear in a few days. If they persist, you can eliminate them with insecticidal soap.

Spider Mites: Spider mites are tiny arthropods related to spiders. Often they deposit fine, barely visible silk fibers on the plant's leaves and stems that resemble tiny spider webs. Nearly always, the largest populations of mites occupy the undersides of leaves. Numerous species of native plants are prone to attack by these pests, especially the carmine spider mite *(Tetranychus cinnibarinus).* Damage can be severe, particularly on *kauila (Alphitonia ponderosa), kokiʻo, nānū, wiliwili,* and some species of *naupaka.* Spider mites suck fluids, often from mature leaves, causing them to mottle, then turn yellow and die prematurely. They also attack the plant's stem tips, resulting in deformed new leaves or stopping their growth completely. The least toxic method of combating spider mites is to vigorously and thoroughly spray water on the plant's leaves, both top and bottom, daily for several days. Usually this treatment physically removes the pest, and the infestation dies out. If it does not, try spraying the plant with insecticidal soap or acephate. When spraying, be certain to spray the stem tips thoroughly in order to reach the spider mites hidden in this tightly packed tissue. Two or three sprayings, one to two days apart, should eliminate the mites.

Stem-Borers

Among the harder-to-detect threats to Hawaiian gardens and to dwindling wild populations of many native trees and shrubs, including rare species such as *uhiuhi* and *mēhamehame (Flueggea neowawraea)*, are small stem-destroying beetles. Perhaps the most notorious of these creatures is the black or coffee twig-borer *(Xylosandrus compactus)* in the family Scolytidae. This alien borer, accidentally introduced to Hawai'i around 1960, attacks over two hundred native and alien species including commercially valuable plants such as coffee and *koa*. (While the black stem-borer is a generalist, other borers in the same family, including some native species, are specialists, attacking only one type of host tree.) An adult female will bore into the stem of a host plant, excavate a chamber, and introduce an ambrosia fungus. She then lays her eggs, which hatch into larval grubs that feed on this fungus. When the larvae mature, they mate and the young females fly away to attack new stems. This whole cycle takes about a month. The boring of the female and the introduced fungus cause the stem to die above the female's initial entry point (Tenbrink and Hara 1994).

In addition to the black stem-borer, we have seen other borers that attack native plants. One species is a fly that lays its eggs on the stems of newly sprouted pea (or bean) family plants, among which is *nanea*. When the eggs hatch, the larvae bore into the stem, killing it in the process. Unfortunately, we have found it hard to positively identify (to species) many stem-borers because we find only the holes they leave behind or perhaps one stage of their life cycle. Nevertheless, in our experience, native plants covered in this book that are most susceptible to stem-borers include *āulu, koa, lama,* and *uhiuhi.* To a lesser degree and under more specific circumstances such as within a nursery, we have also seen *hō'awa, kauila, maua,* and *pāpala kēpau* attacked by borers.

You can sometimes see the early damage caused by stem-borers by inspecting the young branches of a plant at risk. Look for small clumps of sawdustlike material. It brushes away easily and you can then see a small hole where one or more borers have penetrated into the stem. Once an attack is in progress, remedies are labor intensive. You can sometimes kill the borer(s) with a lucky thrust of a pin or by injecting a tiny amount of insecticide into the hole. As a preventative measure, periodic spraying of an insecticide may help protect plants you think will be heavily attacked. Coauthor Culliney used a common commercial "house and garden" insecticide spray on the canopy of a small *lama* tree after several episodes of severe dieback at its new branch tips. (We believe a tiny twig-borer was the cause.) The treatment (light spraying from 2 m; 6½ ft) was repeated several times over a three-month period. During that time, the *lama* tree responded favorably by adding about 15 cm (6 in) of new branches.

Stem-Chewing Pests

A variety of arthropods—insects, sowbugs, and millipedes—and also snails and slugs attack the tender stems of young native plants, especially those that have just been planted in the ground. We have only rarely observed these suspected culprits on or near the plants, so we are not sure which types are actually causing damage or how to rank them in the severity of their attack. In the worst cases, the plants are chewed through or completely girdled (outer sap-containing tissue removed), usually near the ground. Almost always this happens at night.

Fortunately, you do not have to know the identity of the attacker to foil this kind of attack. One technique is to wrap the plant's stem in a small coil of paper. Use scotch tape to secure the wrap, which should extend at least 2–3 cm (1 in) above the ground. A less laborious method is to use the sticky resin, Tanglefoot or Tangletrap. Apply the resin with a toothpick to the base of the stem where it will collect debris and form an effective barrier to chewing pests (see also our section on planting out). Note that this strategy is different from the one we earlier recommended to exclude ants—a sticky resin barrier well above the ground remains fairly clean and thus impassable by ants.

Root-Destroying Pests

The sporadic attacks on some plants by root aphids and mealybugs have already been mentioned. The damage by such creatures is often modest and fairly easily treated with spot applications of chemicals such as Diazinon. In contrast, perhaps the worst underground threats to Hawaiian plants are certain species of nematodes, organisms whose attacks are typically comprehensive, persistent, and eventually devastating.

Nematodes: Most nematodes (also referred to as roundworms or wireworms) are quite small, even microscopic; tens of thousands of them are often present in a single spadeful of soil. They constitute a category of animal that is nearly universally represented in the Earth's biosphere. One zoologist has speculated that if all the rest of Earth's life were to disappear with the nematodes somehow frozen in place, we would see all of the habitats of our planet as ghostly images made of nematodes. Many species live harmlessly in water and soil, but some are parasites in other animals and in plants. One of the worst of these, introduced to Hawai'i as long as 150 years ago, is the root-knot nematode *(Melodogyne* sp.*)*, a threat to numerous plants including agricultural crops and many native Hawaiian species. As the name implies, these nematodes live and multiply in a plant's roots, where they consume nutrients and weaken the plant, eventually killing it. The tiny worms irritate the plant so that the root tissue swells in sites of dense infestations. These swellings, which are often close to the soil's surface, are the root "knots."

Nematode-afflicted plants such as *Achyranthes splendens, uhiuhi,* and *wiliwili* may linger for months or years but always look stressed, with sparse foliage and smaller than usual leaves. Some species such as *kauila (Colubrina oppositifolia)* are killed within a few months of planting out in nematode-infested soil. All show the peculiar gnarled swellings on their roots.

While commercially available chemical treatments for nematodes can be effective for a time in limited areas around plants, you will probably have to continue these toxic applications for the lifetime of your garden. Prevention, whenever possible, is a better strategy. Be careful about importing any bulk soil into an area where you wish to cultivate native plants. One of the coauthors learned this lesson the hard way when his yard became contaminated with root-knot nematodes introduced in a truckload of topsoil delivered by a commercial landscaper. It is far safer to augment soil quality with your own compost or, in small areas, with certified sterile or pest-free products from a reputable nursery. Plant nematode-resistant natives when prevention is impossible. Natives that seem to resist attack by root-knot nematodes include *āulu, hōʻawa, ʻiliahi,* and *maʻo;* it is likely that other resistant species also exist.

Viral, Bacterial, and Fungal Diseases

While information is accumulating rapidly on a variety of pests, especially insects, and their effects on Hawaiian plants, we know much less about native plant diseases. Even the identities of many of the infectious agents on native plants are uncertain. Many of these diseases affect young plants—seedlings in particular. Fortunately, even species that are highly susceptible to infection in the first weeks after sprouting often seem resistant to disease once they have grown past the seedling stage. *Kauila* and *lama* are examples of plants that commonly succumb to rapid wasting diseases when young but whose survivors grow into hardy, beautiful specimens.

The key to success (most of the time) in raising seedlings is to practice good hygiene, water moderately, and avoid crowding. Hygiene refers to the use of clean—or even sterile—soil media for incubating seeds and repotting seedlings. The spores of many disease-causing bacteria and fungi can lie dormant in an old, previously contaminated potting medium. It is always best to use new medium when starting seeds or potting seedlings. Overwatering can promote the disease damping-off, caused by fungi (primarily *Pythium ultimum* and *Rhizoctonia solani*), that results in the sudden death of seedlings. Crowded seedbeds make it easy for the rapid spread of the damping-off fungi and other seed or seedling disease organisms.

A few kinds of plant diseases can be severe in older plants. One of these is the powdery mildew fungus (*Oidium* sp.) that attacks *ʻaʻaliʻi, maʻo, wiliwili,* and some species of *koʻokoʻolau,* often during periods of cloudy weather. Shaded plants also suffer from powdery mildew more often than those out in the open. This fungus produces a filmy white mold that covers the leaves. It may become opaque enough to inhibit photosynthesis, and in the worst infections it causes leaves to die and fall off prematurely. Treat infected plants with sulfur, either as a dry powder or a liquid suspension available from nursery shops. Also, if the plant is potted, try moving it to a sunnier location.

Other diseases, such as those caused by viruses or bacteria, are typically incurable with sprays or powders. However, you can sometimes effectively treat these diseases by cutting away the infected parts of the plant (at least above the ground). Young *hala pepe* frequently exhibit what appears to be an infection that produces necrotic spots on the leaves. Left untreated, the spots spread and can kill a small plant. Treat the disease by cutting off the affected leaves with sharp scissors.

Healthy plants resist many pests and diseases effectively on their own. Often, frequent bouts with pests or diseases are an indication that something is not quite right with the plant's habitat. Perhaps the soil drains poorly or is deficient in important minerals, you are watering too much or too little, or you have planted a sun-loving plant in a shady location. Sometimes altering a plant's environment for the better, such as improving the soil, can be more effective in preventing or combating pests and diseases than any mechanical or chemical "cure."

PROPAGATION AND CARE OF SELECTED SPECIES

The following pages contain our combined knowledge and experiences in the propagation and care of sixty-three species of native Hawaiian plants. We have largely avoided writing on species already addressed by other authors. We have included them only when we felt we could contribute new and significant information on their cultivation. We encourage the reader to seek out and purchase these earlier works (see Literature Cited) because they cover many beautiful and easy-to-grow coastal and dryland plants suitable for many of Hawai'i's gardens.

In our accounts of the individual plants, we present information on cultivating each plant that begins with the collection of mature seeds and ends with the nurturing of healthy plants in a variety of landscapes. We also include information on the natural range and habitat of each plant as it still grows in the wild and, for each, cite how Hawaiians used the plant in construction, toolmaking, dye processing, medicine, and so on. We hope the descriptions and information will help you in your choice of species to grow and give you some ideas on where to plant them.

Yet because we still know so little about the growth of most Hawaiian plants or about their ecology in contemporary habitats, crowded with potentially threatening alien plants and pests, every gardener can find opportunities for experiment and discovery. Communicate your discoveries, successes, and failures with other native plant enthusiasts by joining or becoming involved with one of the many associations, institutions, or societies involved in cultivating and conserving Hawai'i's native plants (for a list of such groups, see Appendix 1 at the end of this book).

For educators at all levels, raising Hawaiian plants offers many opportunities for science experiments and lab exercises. The most reliable information comes from controlled experiments. In such an experiment, a set of plants or seeds is treated in a particular new way (i.e., the experimental group), while another set is kept as controls, which are maintained exactly as the experimental group except for the particular new treatment. If you see differences between the experimental and control groups after the experiment, you can be confident that the treatment caused the change. One difficulty with a controlled experiment is that you need a fairly large number of seeds or plants to avoid misleading results because of individual variations in the plants' responses to the test conditions.

Unfortunately, many of our results reported here are uncontrolled. While many of our findings regarding treatments that speed up sprouting and growth or that protect against seedling diseases are new and intriguing, they should be further studied. We welcome inquiries regarding our databases and methodologies.

Immediately following the name of each selected species in this section, we have included a subjective assessment of how easy that species is to cultivate. There are three categories: *easy, intermediate,* and *difficult.* Easy species germinate readily, often even when shortcuts are made in regard to the techniques prescribed earlier; growth is normally vigorous; and pests and diseases are few or of little consequence to the plant's health. Difficult species present significant challenges and rewards to the grower, either in successfully germinating the seeds or in the subsequent maintenance of the plant. Intermediate species may have one or two difficult aspects related to the plant's cultivation, but these are normally overcome by careful attention to the guidelines we have set forth in this book.

'A'ali'i
(Dodonaea viscosa) — Easy

Habitat

In Hawai'i, *'a'ali'i* is a shrub or small tree that grows in many environments, from coastal areas through dry, mesic and wet forests. It even grows in subalpine shrubland. It occurs on all the main islands except Kaho'olawe. *'A'ali'i* is also native to many tropical areas outside Hawai'i (Wagner et al. 1990).

Hawaiian Uses

Hawaiians used the tough, flexible wood of *'a'ali'i* in the framework of *hale* (houses) and for spears and other weapons. They boiled the seed capsules to produce a red dye for decorating *kapa,* or bark cloth (Degener 1973). Today Hawaiians continue to use the leaves and fruits for leis.

Description

Whether competing with alien plants in eroded pastures and abandoned croplands or inhabiting cold, windswept ridges at the upper edge of Hawai'i's mountain forests, *'a'ali'i* is a survivor. Its attractiveness as well as its hardiness makes it an excellent native species for landscaping. It branches profusely and, with full sunlight and modest moisture, displays a dense foliage of intense green. Its most attractive feature is probably the unusual winged seed capsules that form from spring to fall on female and

Hawaiians used the colorful seed capsules of ʻaʻaliʻi to produce a dye for *kapa* (bark cloth). Today the capsules and leaves are common in leis. ʻAʻaliʻi ranges in stature from a small shrub to a small tree. This one, with white capsules, is quite a tall shrub.

bisexual plants. These delicate, papery capsules turn red or lavender and sometimes yellow or pink, and an ʻaʻaliʻi shrub covered with them is a striking sight. Male ʻaʻaliʻi plants have unusual pollen-producing flowers; Bornhorst (1996) describes them as resembling tiny curled-up octopi. Of course, these male flowers never form the seed capsules. Bisexual ʻaʻaliʻi plants produce both male and female flowers; the latter develop into the winged seed capsules.

Fruits and Seeds

Inside the slightly inflated core sections of each ʻaʻaliʻi capsule are a few tiny (2–3 mm; ⅛ in), dark, spherical seeds that resemble those of cabbage.

Germination and Seedling Growth

Treated briefly in bleach solution and soaked in shallow water overnight, ʻaʻaliʻi seeds sprout readily within a few days to a few weeks. You can decrease the germination time by immersing the seeds in a small volume of very hot water (see previous chapter on hot water scarification). Let the water cool to room temperature and plant the seeds the next day. The small plants develop quickly in vermiculite beds and transplant easily two or three weeks after sprouting. There is some indication that Oʻahu ʻaʻaliʻi, sprouted from seeds collected above Makapuʻu Point, are sensitive to sulfur; these plants were stunted or killed in sulfur-containing potting soil favorable to many other native Hawaiian plants described in this book.

Repotting and Planting Out

Plant out your ʻaʻaliʻi once they are 20–30 cm (8–12 in) tall. They need almost no watering after their first year in the ground and grow well in full sunlight to partial shade. ʻAʻaliʻi thrive in a variety of soils, including those with a high limestone content. Our ʻaʻaliʻi, planted out near sea level on Oʻahu, have grown about half a meter (20 in) per year and produced their first flowers and fruits their second year in the ground. We do not know, but it would be interesting to find out, whether ʻaʻaliʻi sprouted from seed collected from wet forest plants at high elevations would grow well in the dry lowlands and vice versa.

Pests are generally not a serious threat to ʻaʻaliʻi once they are growing vigorously. Scale insects, fostered by ants, sometimes build up on new growth. Spray the affected branches with an insecticidal soap solution or acephate. You may need to repeat this treatment two or three times. Slugs sometimes attack ʻaʻaliʻi at night, consuming foliage up to 2 m (6½ ft) above the ground.

Achyranthes splendens
(no Hawaiian name)

Easy

Habitat

This is a plant of lowland open dry forest or shrubland, most common on rock slopes or coralline plains on Oʻahu, Molokaʻi, Lānaʻi, and Maui (Wagner et al. 1990). One of the two varieties, var. *rotundata*, is federally listed as endangered.

Hawaiian Uses

The Hawaiians do not seem to have used this plant.

Description

Achyranthes splendens is a shrub up to 2 m (6½ ft) tall with silvery leaves. Dense, light-colored hairs produce the silvery color that, in recent times, has given rise to the common name, ʻEwa hinahina. When large, the outer branches of *A. splendens* bend to the ground with the secondary branches growing upward. This gives the plant a somewhat sprawling appearance. *A. splendens* was quite common on Oʻahu's ʻEwa

The silvery appearance of *Achyranthes* has earned it the recently adopted Hawaiian name ʻEwa hinahina. With its many fruits, hundreds on each fruit stalk, you might guess that *Achyranthes* would be a weed—but it is not. With most of its habitat occupied by housing or agriculture, this once common native is now an endangered species.

Plain prior to agricultural development. Now it is restricted to a few fence-enclosed populations in Campbell Industrial Park and Barbers Point Naval Air Station.

Fruits and Seeds

Achyranthes splendens produces a thin-walled fruit called an utricle that contains a single seed. Many small (5 mm; ¼ in) utricles are clustered on long (about 10 cm, 4 in) spikes that normally rise above the body of the plant. They are green when young, light brown when ripe. Look for ripe utricles in the spring and summer. If possible, collect utricles from more than one plant to maintain genetic diversity in your garden. Also, as each plant has hundreds of utricles, resist the temptation to take the entire spike. Instead, cut off part of the spike or collect the numerous utricles by hand, leaving some on the spike.

Germination and Seedling Growth

Plant *Achyranthes splendens* utricles in the standard manner; use many utricles as not all will contain viable seeds. The seeds will begin germinating in two to three weeks. The first leaves are long and thin. Let the seedlings grow several true leaves before separating and transplanting them to individual containers. Gradually, over another two to three weeks, move the seedlings into full sun where they will best develop their silvery leaves. Kept in deep shade, the seedlings will grow spindly stems and abnormally large leaves. Seedlings are occasionally attacked by sucking insects that are easily eliminated with insecticidal soap.

Repotting and Planting Out

Achyranthes splendens responds well to repotting and fertilizer. The plants grow rapidly in pots and after three to six months can be 20–30 cm (8–12 in) tall. Plant out *A. splendens* in full sun and water regularly until established. (Keep potted plants in full sun also.) After outplanting, most plants grow slowly for the first few months, perhaps because the roots need time to grow toward and establish permanent connections to groundwater sources. After this, growth increases and in a year or less the plant can be a meter (3 ft) or so in height and diameter. Reports that this species grows well only in calcareous soils are untrue. The authors have several plants that have been growing vigorously for more than three years in predominantly red clay soil. These plants are even producing seedlings beneath and around them.

A. splendens has a few pests. These include an unidentified cottony mealybug and the small gray weevil *(Myllocerus* sp.*)*. Occasionally, this plant also will be infested by scale insects, but this is usually only true of unhealthy plants or when ants introduce and protect them. Eliminate the scale with two or three treatments of insecticidal soap or acephate. Use an ant bait insecticide or Diazinon to get rid of the ants. A more serious pest is an unidentified stem-borer. Infestation by this insect can be overlooked because the stem is not always killed. Look for holes 1–3 mm (1/16 in) in diameter in the stems as evidence of this pest. Infested plants also have fewer and yellowing leaves. In the past, the authors have treated this problem by either removing the infected stems or destroying the entire plant, depending on the degree of infestation. We have not tried any type of chemical treatment.

'Āheahea or *'Aweoweo* (*Chenopodium oahuense*)   Easy

Habitat

'Āheahea grows primarily in coastal and lowland shrubland and forest. It is also found in subalpine shrubland (Wagner et al. 1990). *'Āheahea* grows on Lisianski, Laysan, French Frigate Shoals, Necker, Nihoa, and all the main islands except Kaho'olawe.

Hawaiian Uses

Hawaiians ate *'āheahea* leaves when food was scarce and used the wood for composite fish hooks (Krauss 1993).

Description

'Āheahea is a member of the goosefoot family and is related to the common lambs quarters *(Chenopodium album* L.*)*. But unlike most species in the genus *Chenopodium*, *'āheahea* is a woody shrub or occasionally even a small tree. In fact, *'āheahea*

'Āheahea fruit stalks turn light brown when mature. This young *'āheahea* just appeared one day in one of our gardens. The seed probably came from a mature plant that lived and died (after about three years) a few meters away.

is the tallest and woodiest species in the genus. This is not an unusual attribute for native Hawaiian plants. The evolution of small nonwoody plants into woody shrubs and even trees, referred to as arborescence, has occurred repeatedly in Hawai'i. Examples include 'akoko, kulu'ī, na'ena'e, pāpala, the Hawaiian lobeloids, and the Hawaiian violet. There are several proposed hypotheses to explain this evolutionary trend. They include an ecological shift toward wetter and wetter forest species, a moderate climate that promotes elongation of stems, the evolution of larger seeds, and others (see Carlquist 1980).

'Āheahea has small, attractively lobed leaves, somewhat reminiscent of maple leaves but smaller. Flowering heads are numerous and nearly a constant feature of the plant. In cultivation, 'āheahea is a short-lived perennial. None of the plants we have planted out have lasted more than four years. After about three years, the plants begin to put out fewer leaves and some branches die back to the main trunk. We have seen similar aging in wild plants. Hopefully this is true only for the O'ahu variety that we have grown; perhaps other varieties live longer.

Fruits and Seeds

The small (2 mm; 1/16 in) seed-bearing utricles of 'āheahea are clustered at the end of branches. As utricles ripen, the stem bearing them dies and the stem and utricles turn from green to light brown. The utricles come off easily when collected by hand.

Germination and Seedling Growth

Plant 'āheahea utricles the standard way. Because the utricles are small, however, do not bury them deep in the vermiculite and use only a light covering of green moss or none at all. The seeds begin germinating in one week and continue to germinate for two to three more. Seedlings put on their first and second set of true leaves in less than a month. After the second or third set of leaves, transplant the young seedlings to individual pots. The seedlings will be 20–30 cm (8–12 in) tall and ready for planting out in two to three months.

Repotting and Planting Out

Plant out 'āheahea using the standard guidelines in the previous section. 'Āheahea does not seem to be particular about soil type and once established, after two or three months, should not require supplementary watering. In less than a year, your 'āheahea should be nearly full size, 1–2 m (3–6½ ft) tall. You may wish to prune the shrub as branches occasionally die back to the main trunk.

On two occasions, after particularly heavy rains, we have had mature 'āheahea shrubs die back to one or two main stems. This suggests to us that the plant may not be able to tolerate extremely wet soils.

We have encountered few pests on *'āheahea*. The most frequent is the small gray weevil *(Myllocerus* sp.*)*, which eats the leaves. On most occasions, we have ignored the weevils and they eventually disappear. (We have seen the local birds hunt these weevils on the *'āheahea*.) Occasionally, aphids and mealybugs infest the stem tips of *'āheahea*. The weevil, aphids, and mealybugs are easily controlled by either insecticidal soap or acephate.

'Āheahea is one of the few native plants we have grown that can successfully germinate and grow amidst the alien vegetation in Hawai'i's lowland environment. Other species include *'a'ali'i*, *'ilima*, and *ma'o*. On numerous occasions after heavy rains, we have had new *'āheahea* sprout and grow up around existing plants. The farthest seedling was over 3 m (10 ft) away! These new plants typically start in a small bare patch of soil, surrounding or underneath the parent. (We have not seen *'āheahea* sprout in thick grass.) Once established, with three or four sets of true leaves, they have rarely died and usually grow into mature plants. *'Āheahea*'s ability to reseed and its tenacity in alien shrublands may make it a useful species for those attempting native restoration in Hawai'i's lowlands.

'Aiea
(Nothocestrum breviflorum) Difficult

Habitat

This small, rare tree occurs on dry slopes from (formerly) near sea level to about 1,800 m (6,000 ft). There are four species of the endemic genus *Nothocestrum* collectively found throughout the main islands (except Ni'ihau and Kaho'olawe). Two of the four species are federally listed as endangered. The only significant wild population of *N. breviflorum* is on ranch lands in the Big Island's North Kona District. But year after year, brush fires take a toll on these small trees, and they are not reseeding themselves. *'Aiea*'s vulnerability is largely due to the dense growth of alien grasses in its dry leeward habitat. The grasses present a double threat: They fuel extremely hot fires, and they are thought to prevent the growth of tree seedlings.

Hawaiian Uses

Hawaiians used the soft, light-colored wood of *'aiea* to fashion gunwales for canoes (Krauss 1993). They also may have used some part of the plant as medicine (Lamb 1981).

Description

'Aiea is unusual in being a tree member of the nightshade family (related to tomato and eggplant). Endemic to Hawai'i at the generic level, *'aiea* has become exceedingly

Once pollinated, the tubular, yellow flowers of *'aiea* will develop into orange fruits. Hawai'i's ugliest tree? Perhaps, but how often does a tomato grow into a tree? (*'Aiea* is related to the tomato and eggplant.)

rare and prone to extinction. Field botanist Joseph Rock, who cataloged Hawai'i's native trees early in the twentieth century, called *'aiea* Hawai'i's ugliest tree (Rock 1974). Hawai'i gardeners who raise this endangered species may judge for themselves. *'Aiea* has twisted branches, rough bark, bright green and satiny foliage, dense clusters of small, tubular, greenish yellow blossoms, and orange, berrylike fruits. Some may disagree with Rock. Indeed, bonsai enthusiasts may be attracted to this species for its naturally contorted form and for its rarity as one of the world's few woody nightshades.

Fruits and Seeds

'Aiea fruits ripen from spring to summer. They average about the size of a pea, are orange when mature, and contain up to ten small seeds that resemble those of a tomato.

Germination and Seedling Growth

After standard preparation (do not expose the seeds to a dilute bleach solution for more than ten minutes), *'aiea* seeds sprout in two weeks to one month. The sprouts closely resemble eggplant seedlings. *'Aiea* seedlings grow at a moderate pace, reaching

10–15 cm (4–6 in) in height after three months. Aphids and red mites occasionally infest the undersides of the leaves. On very young plants we suggest removing these pests by hand; a moist cotton swab or piece of paper towel works well on larger leaves. Older seedlings will tolerate spraying with an insecticidal soap solution.

Young *'aiea* often show spurts of growth followed by dormant periods. This pattern is also common in other Hawaiian species we have grown such as *lama*, *kōlea*, and *kōpiko*, but it may be more pronounced in *'aiea*. Resist the temptation to heavily fertilize *'aiea*; they seem to shock easily and will lose all or most of their leaves. Sometimes the plants recover; sometimes they die.

Repotting and Planting Out

Shock may also occur after repotting or planting out, but its effects seem to be lessened by first watering with a dilute vitamin B-1 solution made from a concentrate such as Dexol, available at garden shops. If your plant still loses its leaves after repotting or planting out, we recommend against excessive watering. The plant usually recovers on its own with time.

Near the ground, *'aiea* are sometimes attacked by chewing pests. Millipedes, sowbugs, and slugs are probable culprits. The young bark, which is soft, does not protect against this attack even in plants up to 30 cm (1 ft) tall, which can be girdled and killed in one or two nights. We have used a coat of Tangletrap or Tanglefoot resin on the lower stem, extending a centimeter or so below the soil line, to protect against these pests. If the attack on stems and shallow roots is especially severe, try periodic applications of granular Diazinon to kill the surrounding pests.

Young, healthy *'aiea,* planted out in full sun, can grow as much as 30 cm (1 ft) per year. These older plants, like seedlings, tend to undergo spurts of growth with intervening dormant periods. In a number of cases, the apical or leading branches of our plants have wilted and died for unknown reasons. Such a partial die-back usually triggers a burst of new growth nearer to the plant's base. This phenomenon is similar to that seen in *maiapilo (Capparis sandwichiana)*. While *'aiea* may be classified as a tree, in our experience it commonly seems to be trying to remain a shrub.

'Ākia
(Wikstroemia oahuensis, W. phillyreifolia, W. sandwicensis, and W. uva-ursi) — Easy

Habitat

Early Hawaiians recognized the numerous species of *Wikstroemia*, growing from near sea level to over 2,000 m (6,600 ft), as a closely related group of plants. Today a dozen species of *'ākia* are recognized by Hawai'i's botanists. All twelve are thought to have evolved from a single colonist that spread and adapted to a variety of habitats in Hawai'i. These habitats include dry rocky coasts, barely vegetated lava fields, hot lowland dry forests, cool upland wet forests, and soggy bogs.

W. oahuensis is the most widely distributed of the twelve 'ākia species and is found on all the main islands except Hawai'i, Kaho'olawe, and Ni'ihau. It is usually a shrub but in some wild places classifies as a small tree (maximum height of about 6 m; 20 ft). It has adapted to mostly mesic and wet forest conditions.

W. phillyreifolia is a small-leafed, long-branched shrub with small, bright red fruits that grows on young lava soils to disturbed pasturelands in mesic habitat up to about 2,000 m (6,600 ft) on the Big Island. People sometimes mistake this plant for *'ōhelo (Vaccinium reticulatum),* the red-berried Hawaiian relative of the mainland blueberry. The two, *W. phillyreifolia* and *'ōhelo,* grow very well together in suburban gardens at higher elevations such as the cloudy, wet regions of Volcano on the Big Island.

The large (and large-leafed) 'ākia species, *Wikstroemia sandwicensis,* was common in the native dry coastal forest of western Puna on the Big Island—a community that has nearly vanished during more than a decade of recent lava flows from Kīlauea's East Rift. Near Kamoamoa on the coast of Hawai'i Volcanoes National Park, this species thrived on the young volcanic terrain, reaching the stature of a small tree.

W. uva-ursi, the smallest of the four species we have raised, has round fruits and leaves. It is mostly a coastal or dryland 'ākia, now quite popular with landscapers who use it in dense border plantings.

At least one 'ākia, the low-growing *Wikstroemia uva-ursi* (left), has already become a favorite native plant of amateur and professional landscapers. Remember, however, that because some 'ākia have unisexual flowers such as this *W. sandwicensis* (right), you will need to have at least one male plant and one female plant in your garden to enjoy seeing the colorful orange or red fruits.

Hawaiian Uses

'Ākia bark, fibrous and strong, was an important source of cordage in ancient Hawai'i (Stone and Pratt 1994). Hawaiians also used *'ākia* as a laxative and for treating asthma (Wagner et al. 1990). At least some *'ākia* are highly toxic to vertebrate animals. Hawaiians made use of this property by placing handfuls of pulped *'ākia* leaves and stems, wrapped in coconut fiber, into shoreline pools to stun fish, a method called *hola*. Fish caught or killed in this manner were, however, safe to eat. Hawaiians also made a drink from these poisonous *'ākia* that they used for suicide and the execution of criminals (Degener 1973).

Description

All *'ākia* species are either shrubs or small trees. Based on their habitats and our own experiences, the species we have raised show wide tolerances in environmental factors such as temperature, rainfall, and soil type. *'Ākia* bark has a distinctive ringed pattern. Combined with their normally abundant leaves and plentiful small, bright red, orange, or yellow fruits (that appeal to birds), all the *'ākia* species are attractive plants. During much of the year, they produce clusters of tiny long-throated yellow flowers with a fragrance resembling honeysuckle. In at least some species, this fragrance is only slightly noticeable during the day but greatly intensifies around dusk, suggesting that these plants attract moths as pollinators.

It has long been thought that most Hawaiian *'ākia* are less poisonous than their Indo-Pacific relatives, having lost their toxicity in an evolutionary setting that lacked browsing mammals (Carlquist 1980). Obviously, however, certain species and varieties have remained highly toxic, or we would not have the historical references to uses of *'ākia* to stupefy fishes and kill humans. Because of the uncertainties regarding toxicity in individual plants and the possibility of cross-breeding among nominal species, it is probably best not to grow any *'ākia* in areas frequented by young children.

Fruits and Seeds

The most reliable time to find ripe, colorful *'ākia* fruit is fall and winter. Under cultivation, however, fruiting often occurs at intervals throughout the year. Each fruit has a single ovoid pit (the seed) at the center. Small bits of pulp left clinging to the seeds can be safely ignored after the standard bleach treatment.

Germination and Seedling Growth

Plant *'ākia* seeds using our standard method. The seeds are generally resistant to fungus. Sprouting often takes a month or more, especially in xeric species. However, once the first seeds begin to sprout, the others in the tray are quick to follow. Transplant young seedlings to individual pots after two or three sets of true leaves appear. Seedlings are hardy and resistant to fungal rot and damping-off diseases. They are also largely immune to insect attack. Initial growth is slow but accelerates after a few months.

Repotting and Planting Out

Ākia transplant and survive well in both a pot and in the ground. Two plants of *Wikstroemia sandwicensis*, raised in Waimānalo, Oʻahu, and planted out on calcareous sandy soil, look no different from their wild ancestors on the Big Island. One of these is now 4 m (13 ft) tall after six years. Growth is proportionately vigorous in the other three species of *ʻākia* we have grown. Within about a year of planting out, you can expect to see your *ʻākia* develop flowers and fruits. Note, however, that some develop only "male" or only "female" flowers. Plants with male flowers will not develop fruit.

The *ʻākia* we have grown have been only rarely attacked by insects or other pests. On occasion, we have seen a plant with a smooth semicircular bite taken from several of its leaves. This is the work of a leaf-cutting bee. Snails and slugs will sometimes attack and kill small, newly planted *ʻākia*.

ʻAkoko
(*Chamaesyce celastroides, C. degeneri,* and *C. skottsbergii*)   Intermediate

Habitat

Chamaesyce celastroides is found from the coast to upland mesic forest on all the main islands and Nihoa. One variety, var. *kaenana*, found at Kaʻena Point, Oʻahu, is federally listed as endangered. *C. degeneri* grows in coastal areas on all the main islands except Lānaʻi and Kahoʻolawe. *C. skottsbergii* is found in coastal areas and dry shrubland on Oʻahu, Molokaʻi, Maui, and Kahoʻolawe. One variety, var. *skottsbergii*, found on Oʻahu and Molokaʻi, is federally listed as endangered (Wagner et al. 1990).

Hawaiian Uses

The milky sap of some *ʻakoko* was one ingredient of the paint used on canoe hulls (Krauss 1993).

Description

The *ʻakoko* (there are about fifteen endemic species in Hawaiʻi) are an excellent example of adaptive radiation (evolution). Carlquist (1980) suggests that the founder

'akoko

This attractive endangered *'akoko, Chamaesyce celastroides* var. *kaenana*, grows wild only on the rocky coastal slopes at Ka'ena Point, O'ahu. Fortunately, this spreading shrub is not difficult to grow and cultivated plants can be seen in several O'ahu gardens. Nearby, a smaller relative, *C. degeneri,* is great for a sand-filled planter on a sunny lanai. Damaged or old leaves turn a blood red as they die.

plant to Hawai'i was probably a coastal species similar to *Chamaesyce celastroides* that even today shows considerable variation in morphology (from upright bushes to wide, spreading, matlike shrubs). From this ancestor, new species evolved that were better adapted to the wetter forests as one goes up Hawai'i's mountains. These upland species developed less papery leaves that in dryland species fall off during times of drought, and they grew taller to catch the limited sunlight in thick forests. (Evolving taller forms is a phenomenon known as arborescence, frequently seen in island species.) Meanwhile, lowland and coastal species such as *C. degeneri* and *C. skottsbergii* evolved the other way. They became increasingly deciduous and prostrate. (Some botanists believe that these coastal species evolved from a second more coast-adapted founder.) Both *C. degeneri* and *C. skottsbergii* are rather diminutive plants, *C. degeneri* being the smallest, no more than 25 cm (10 in) tall and about the same in diameter. *C. celastroides* is a larger shrub, at most 1–2 m (3–6½ ft) in height and diameter.

Fruits and Seeds

The flowers on all *'akoko* are small, a few millimeters in diameter, as are the fruits, sometimes called capsules. Within each capsule are two or three small seeds; brown to gray in *Chamaesyce celastroides* and *C. skottsbergii,* white in *C. degeneri*. All three species flower and fruit repeatedly throughout the year. It is best to collect seed just before or just after the capsules split open; you are then sure to have fresh but mature seeds.

Germination and Seedling Growth

Plant *'akoko* seeds in the standard manner. Because of the small size of the seedlings, however, we suggest placing little or no green moss over the seedbed. Take extra care to avoid letting the seedbed dry out. The seeds will begin sprouting in one to two weeks (*Chamaesyce celastroides* sometimes take longer). Let the seedlings grow to about 5 cm (2 in) in height before separating and transplanting to individual containers.

You can also propagate *C. celastroides* and *C. degeneri* by stem-tip cuttings. (We have not tried to propagate *C. skottsbergii* from stem tips.) Our success from cuttings is about 50 percent. Take cuttings about 10 cm (4 in) in length from actively growing tips with mostly green wood. Use the standard procedure for stem-tip cuttings described in the last section. Rooting time is variable; as little as one week in *C. degeneri* or as long as two to three months in both species. After root growth and new stem growth are obvious, transfer the new plant to a standard potting mix.

Repotting and Planting Out

We have had no problems in repotting ʻakoko. To date, we have planted out only *Chamaesyce skottsbergii*. Most have done well, growing to half a meter (about 2 ft) in less than a year and flowering and fruiting profusely. Of the eight plants, two died in less than a year, perhaps because we watered them too much. All of the plantings were in noncalcareous soil, in contrast to where this species is naturally found.

Our outplantings of *C. celastroides* and *C. degeneri* have been limited to large planters filled with beach sand, coral rubble, and cinder with a small amount of red clay soil mixed in. Survival and growth over three years in these containers has been good. The *C. degeneri* reached mature size in less than a year. Since then we have repeatedly pruned this species to encourage new growth. All three species of ʻakoko respond well to liquid fertilizer (once every one to two months).

Pests of ʻakoko include whitefly, scale insects, and mealybugs. Often the scale insects and mealybugs are brought in and protected by ants. All three pests can be eliminated with two or three treatments of insecticidal soap or acephate. Of the three pests, the whitefly is least threatening to the plant's survival.

ʻĀlaʻa
(*Pouteria sandwicensis*)   Intermediate

Habitat

This tree occurs from about 200 to over 1,000 m (660–3,300 ft) on all the main islands except Kahoʻolawe and Niʻihau. It grows in dry to mesic forest (Wagner et al. 1990).

Hawaiian Uses

Hawaiians used the hard wood of ʻālaʻa in housing, for ʻōʻō (digging sticks), and spears (Lamb 1981). ʻĀlaʻa was a birdlime tree; Hawaiians used its sticky sap to capture small birds for their colorful feathers (Wagner et al. 1990).

The ʻālaʻa has fruits the size of golf balls that ripen in fall. On some islands, such as here on Hawaiʻi, the tree produces yellow orange fruits. On others, such as Oʻahu, the fruits are dark purple. Sometimes you will see ripe but smaller marble-sized fruits. These normally do not contain viable seed.

Description

Formerly classified in the genus *Planchonella,* this is a small to medium-sized tree reaching about 15 m (50 ft) in height. However, in the dry forests of Kānepuʻu, Lānaʻi, ʻālaʻa is considerably shorter than elsewhere, the largest trees being about 5 m (17 ft) tall. For mainlanders, its rounded shape is somewhat reminiscent of an apple tree.

ʻĀlaʻa is quite variable in appearance. For example, in addition to their smaller stature, ʻālaʻa trees on Lānaʻi have leaves about 8 cm (3 in) long with many brown hairs on their lower surfaces. This gives the entire tree a rusty color. In contrast, on Oʻahu, the leaves are at least twice as long and have little hair; the foliage is a more typical green color. On Lānaʻi and Hawaiʻi, ʻālaʻa fruits are bright yellow orange when ripe. On Oʻahu, the fruits are dark purple.

Fruits and Seeds

Beginning in late fall, ʻālaʻa decorates itself with fruits the size of golf balls that turn from green to either purple, orange, or yellow when ripe. Collect only ripe or ripening (not growing) fruits. You can test green fruits to see if they are almost ripe by gently squeezing them. Unripe, still growing fruits are very hard and you should leave them on the tree (the seeds inside will not sprout). In contrast, you can pick slightly soft fruits and ripen them in a paper bag; this takes one to two weeks. Additionally, viable seeds are often on the ground under a tree. Inside the fruit are several large seeds. They have a thick seed coat, so germination is aided by fairly deep scarification.

Germination and Seedling Growth

'Āla'a seeds sprout in one to two months using our standard method. Fungi or bacteria sometimes attack and kill 'āla'a seeds while they are incubating on seedbeds. Test them by squeezing between your (clean) fingers. Healthy seeds are hard; diseased seeds turn to mush and crumple. You should discard diseased seeds promptly to prevent the fungus from spreading to healthy seeds. If the problem is widespread, consider using a fungicide on the seeds before sowing.

After a crack develops around the narrow base of the seed, a stout white root emerges. A week later, the thick cotyledons shrug off the remnants of the husk as the seedling becomes erect. Seedlings are robust and generally disease resistant. Even so, they are sometimes attacked at the stem tip and killed, apparently by a tiny insect that also attacks a variety of other seedlings such as *lama* and *pāpala kēpau*.

Repotting and Planting Out

Seedlings quickly develop a long, wiry tap root resembling that of *lama*. While small 'āla'a survive transplanting well, do not use a shallow pot as the tap root will begin to coil extensively. 'Āla'a seedlings grow slowly for up to several months but sometimes grow more rapidly thereafter. In general, however, it seems they do not like living in a pot and young plants should probably be planted out sooner rather than later. Unfortunately, we have failed to follow our own advice and have yet to plant out our 'āla'a, and thus we cannot comment on later growth or pest problems. Coauthor Koebele has observed that trees in the wild, such as those growing at Kānepu'u on Lāna'i and near Pu'uwa'awa'a Ranch on the Big Island, show little evidence of attack by insect pests. Hopefully this will prove true for the widespread cultivation of this beautiful tree.

Alahe'e
(*Psydrax odorata*) Intermediate

Habitat

Alahe'e is found from near sea level in dry shrubland to mesic and perhaps even wet mountain forest on all the main islands except Ni'ihau and Kaho'olawe (Wagner et al. 1990).

Hawaiian Uses

Native Hawaiians fashioned fishhooks, light spears, and a variety of tools from the hard wood of *alahe'e*. They also produced a black dye from the leaves (Krauss 1993).

Alahe'e, with its dense flower clusters, shiny foliage, and pleasant "slippery" fragrance, is one of the most desirable small trees for a Hawaiian garden.

Description

Alahe'e is a small tree or shrub with glossy dark green leaves and fragrant flowers that betray its membership in the coffee family of plants (Rubiaceae). The flowers are small and white, occur in clusters, and their fragrance is elusive. Directly beneath your nose the flowers' fragrance is weak, but if you stand downwind of an *alahe'e*, the sweet scent is strong and pleasant. The Hawaiians characterized this in the plant's name: *'ala* means "fragrant" and *he'e* means "slippery" or "slippery like an octopus." *Alahe'e*, therefore, is the plant with the slippery fragrance (Bornhorst 1996).

Alahe'e is indigenous to Hawai'i. Although it arrived in Hawai'i unassisted by humans, it is also found on many other islands throughout the South Pacific (i.e., it is not endemic to Hawai'i). *Alahe'e* is still quite common in Hawai'i's forests but, like nearly all our native plants, is becoming less so. This is unfortunate; it is an attractive plant that has not escaped the eye of professional horticulturists. Unfortunately, while *alahe'e* is easy to grow from seed (see below), quite often all the seeds on wild plants are destroyed by a moth *(Orneodes objurgatella)* that lays its eggs on the fruit. The eggs hatch and the larvae eat the seeds. Some Hawaiian entomologists, such as Dr. Steve Montgomery (pers. comm. 1998), think the moth is a native species; others (Zimmerman 1958) believe the moth was introduced. Occasionally, *alahe'e* escape detection by this pest and you can find good seed. Repeated sprayings of insecticide during the fruiting period may also keep this insect at bay (Bornhorst 1996). Propagation from stem-tip cuttings might be a detour around this problem except that until recently no one had successfully propagated *alahe'e* from cuttings. In 1997, however, Dr. Richard Criley, a professor in the Department of Horticulture at the University of Hawai'i, was able to prompt about 10 percent of nine hundred *alahe'e* cuttings to root (pers. comm.). While Dr. Criley has not worked out all the details, he suggests using green wood cuttings, a strong rooting hormone treatment, and patience. (Rooting in his experiments took ten to twelve weeks.)

Fruits and Seeds

The *alahe'e* fruit is a berry, about 1 cm (⅜ in) in diameter, normally containing two seeds but occasionally only one. The seeds have a hard, rough coat and a half-moon shape. Collect the ripe, soft, black fruits in late fall or winter. Unripe fruits are green and, from our experience, do not contain seeds capable of germination. Examine each fruit carefully for one or more holes. This is evidence of attack by the moth. These infected fruits nearly always contain inviable seeds.

Germination and Seedling Growth

Plant *alahe'e* seeds in the standard manner. Germination takes approximately one month and is very reliable with fresh, undamaged seed. The seedling's first leaves (the cotyledons) are long and thin; the second true leaves resemble those of the mature plant. Transplant the seedlings from the germination tray into individual pots after they develop their first or second pair of true leaves. About one month after transplanting, begin watering the plants with a liquid fertilizer (at half strength) or add a small amount of time-release fertilizer pellets. About two months after transplanting, seedling growth seems to increase and remain rapid. After four to six months, seedlings can be 20–30 cm (8–12 in) tall and ready for repotting or planting out. Seedlings (and more mature plants) are sometimes plagued by scale insects. Remove these by hand or treat several times with an insecticide such as acephate.

Repotting and Planting Out

Alahe'e survive repotting and planting out well. Still, for the first couple of months after outplanting, examine them carefully for signs of drought stress; these are wilting leaves and stems followed by leaf loss. Growth of *alahe'e* after planting out seems largely dependent on the amount of water and fertilizer it receives—with some moderation, of course. Unfortunately, well-fertilized as well as drought-stressed plants appear to be more prone to infestation by scale insects or—occasionally—mealybugs. Often too, the scale is introduced and protected from predators by ants. Eliminating these pests can sometimes be a protracted battle, involving repeated removal by hand (on seedlings) or spraying of insecticides, such as insecticidal soap or acephate. However, it is a battle that must be won because severe infestations of scale insects can kill the plant. Healthy, well-watered, and well-fertilized *alahe'e* will grow about half a meter (1–2 ft) per year. In drier locations, growth is slower. While adapted to dry shrubland and forest, our cultivated *alahe'e* appears less capable than other plants, such as *naio, 'ilima,* or *ma'o,* to survive drought. For this reason, continue to keep a close eye on your *alahe'e* even several years after outplanting if you live in a particularly dry area (e.g., less than 76 cm; 30 in of rain per year).

Āulu and *Mānele*
(Sapindus oahuensis and *S. saponaria)*   Easy

Habitat

Āulu occurs in mesic to dry forest on Kaua'i and on O'ahu in the Wai'anae Range and from Waimalu to Niu in the Ko'olau Range. *Mānele* occurs in Mexico, South America, the Pacific basin, and Africa. In Hawai'i, it grows wild in mesic forest on the Big Island (Wagner et al. 1990).

Hawaiian Uses

Hawaiians used *āulu* seeds medicinally as a cathartic and for leis; the wood was used to make spears and for house building (Lamb 1980).

Description

Large *āulu* trees can be 15 m (50 ft) tall. The species is unique because it has simple leaves; the other species in the genus have compound leaves. *Mānele* is a deciduous

Ripe *āulu* fruits look, feel, and smell like figs but you would not want to eat them! When mixed with water, the fruit of these Hawaiian soapberry trees makes a sudsy lather. Both *āulu* (left) and the *mānele* (right) grow quickly and make attractive shade trees.

tree up to 25 m (80 ft) in height. *Āulu* and *mānele* are both soapberry trees. Mixed with water, *āulu* or *mānele* fruit pulp produces a soapy lather. In some places, such as India and Mexico, people wash their hair and delicate fabrics with soapberry (Neal, 1965). With its large leaves, *āulu* is one of the few Hawaiian dryland trees that produces a significant shade. Perhaps because of this it will find a new home in our often hot and sunny cities.

Fruits and Seeds

Āulu fruits are 2–3 cm (1 in) long and resemble figs when ripe. Each fruit contains one large, black, ovoid seed with a rough texture. *Mānele* fruits are oval, about 2 cm (¾ in) in diameter, and brown to black when ripe; they contain one black, oval seed. Collect the mature fruits during the fall and winter, either from the tree or the ground beneath. Immature fruits do not ripen properly off the tree. After removing the seed from the fruit, inspect it carefully for insect damage. Discard any seed that has even the smallest hole. Seeds that float in tap water are usually inviable.

Germination and Seedling Growth

An experiment conducted by coauthor Koebele and his students revealed that, with the seed coat intact, germination of *āulu* and *mānele* was extremely poor in a mixture of fine cinder and potting soil. After two months, none of the seeds had germinated. On closer inspection, many revealed damage by small insect larvae or bacteria that had turned the embryo to mush. Germination was equally poor when the seed coat was cracked. In contrast, when the outer black and inner brown seed coats were carefully removed, germination was very good and very rapid (within one week).

To remove the seed coats, we suggest clipping the outer coat with garden clippers (or a knife) and placing the seed in new, moist vermiculite for one to two weeks. This will soften the outer seed coat, permitting you to remove it by hand or with a small knife. Be extremely careful removing the thin inner papery brown seed coat because at one point it wraps under the embryonic root. Without care (or practice) you will break the root while removing this seed coat. After removing both seed coats, plant the embryo immediately about 2 cm (¾ in) deep in clean, moist potting soil.

Seedling survival is excellent and growth is rapid. We have found that some *āulu* embryos sprout with yellow or unusually thin and distorted leaves. Our guess is that this is a genetic defect (albino seedlings?) and not due to any pest or disease. We recommend discarding these plants.

Repotting and Planting Out

Āulu and *mānele* tolerate transplanting well and grow quickly when fertilized. After one year, with good care and moderate fertilization, your *āulu* will be a meter (3 ft) tall. *Mānele* grows a little more slowly. One *āulu*, planted out three years ago, is now over 3 m (10 ft) tall. Some botanists say the Wai'anae and Ko'olau races of *āulu* are morphologically distinct (i.e., they look different), so the two should probably be kept separate when outplanted. We have not propagated the *āulu* from Kaua'i.

In nonwindy locations, whiteflies occasionally attack *āulu* but do not cause any real damage. In contrast, a small gray weevil *(Myllocerus* species, see previous section*)* can significantly damage young leaves. Keep the weevil under control with frequent sprayings of insecticidal soap or acephate (e.g., Isotox at 2 tbs per gallon). Chinese rose beetles *(Adoretus sinicus)* can also be a problem. Curiously, the endemic *āulu* is more susceptible to weevil and Chinese rose beetle damage than is the indigenous *mānele*.

On a couple of outplanted trees, we sometimes see damage from twig borers, both the black or coffee twig borer *(Xylosandrus compactus)* and another unidentified and larger larval insect. Both borers prefer young branches. After the borer hollows out the stem, the branch's leaves wilt and turn brown. Quickly remove and destroy the damaged branches with the borers inside. Since the borers' preference is for young branches, older trees are less likely to receive a fatal hit and, to date, none of our plants has been killed by borers.

'Āwikiwiki
(Canavalia galeata) Easy

Habitat

This species of *'āwikiwiki* is a vine that grows in mesic shrubland or forest at modest elevations of a few hundred meters. In the wild, it occurs only on O'ahu. Seven related species, both native and introduced, occur in coastal, dryland, and mesic regions throughout the main islands (Wagner et al. 1990).

Hawaiian Uses

Hawaiians made fishnets and traps from the tough flexible stems of *'āwikiwiki* (Abbott 1992). They also used it as an ingredient in a lotion applied to skin disorders (Krauss 1993). Today, as in the past, *'āwikiwiki* flowers are used in lei making.

Description

A many-branching vine adapted to dry or mesic conditions, *'āwikiwiki* makes an attractive cover for unsightly fences or a more traditional trellis. Each leaf is composed of three glossy leaflets that are slightly reddish when young. If you want a more natural appearance, plant *'āwikiwiki* near a tree in your yard so it can climb as it normally does in its Wai'anae Range home. It can even be used as a ground cover on expansive rockscapes. Its most attractive feature is its large purple flowers, like giant sweet peas, with white streaks. These flowers can be fashioned into a unique and rare lei.

This wild ʻāwikiwiki climbs up a *koa* tree. At home you can have it do the same or let it cover a chainlink fence. The pealike flowers of ʻāwikiwiki are sometimes used in leis; they develop into large flat pods.

Canavalia galeata, as mentioned, naturally occurs only on Oʻahu, but there are five other native species of ʻāwikiwiki scattered throughout the Hawaiian Islands. Some grow on several of the main islands, such as *C. hawaiiensis,* found on Hawaiʻi, Lānaʻi, and Maui; others are restricted to a single island, such as *C. kauaiensis.* While we have not grown the other ʻāwikiwiki species, we encourage neighbor island gardeners to do so. Each has flowers that rival the beauty of Oʻahu's species, and each, we suspect, will be equally easy to grow.

Fruits and Seeds

ʻĀwikiwiki is in the bean family of plants (Fabaceae). It forms large pods about 10 cm (4 in) long. Look for the ripe brown pods in either fall or winter. You will find several large, reddish brown, flattened bean-shaped seeds within each pod.

Germination and Seedling Growth

You should scarify ʻāwikiwiki seeds to ensure rapid sprouting. Soak the scarified seeds in water overnight before planting. The seeds will swell noticeably. After sowing, a stout white root appears within a week, followed rapidly by the first set of true leaves that emerge from between the fleshy cotyledons. At this stage, remove the seedlings from the seedbed tray and plant them in individual pots. An alternative and more direct planting method is simply to place the scarified, water-soaked seed directly into a pot containing a moist, clean potting medium. While you are waiting for the seed to sprout, water the pot to keep the potting mix moist.

Repotting and Planting Out

Young *'āwikiwiki* grow very rapidly from the start, doubling in size each week or two. They branch quickly into two or three spreading vines that begin climbing any nearby support, including neighboring plants. *Āwikiwiki* is sometimes attacked by scale insects and the black stinkbug, a potentially serious pest (see previous section for a complete description of this pest and how to combat it).

Hala pepe
(Pleomele forbesii, P. halapepe,
and *P. hawaiiensis)* Intermediate

Habitat

There are six native species of *hala pepe* in the Hawaiian Islands. They occur from 120 m (400 ft) to 1,200 m (4,000 ft) elevation in either dry or mesic forest or both. *Pleomele forbesii* and *P. halapepe* occur naturally only on O'ahu; *P. hawaiiensis* only on Hawai'i (Wagner et al. 1990). *P. hawaiiensis* is federally listed as endangered.

This young *hala pepe* at Amy B. H. Greenwell Ethnobotanical Garden is already producing an abundance of fruit. Looking like a cluster of cherries, *hala pepe* fruits usually ripen in summer or fall.

Hawaiian Uses

Hawaiians used *hala pepe* flowers for lei making (Krauss 1993). Inside a *hālau hula*, a building for dancers and trainees, a branch of *hala pepe* was placed on the *kuahu* (altar) to represent the goddess Kapo (Abbott 1992). Hawaiians also carved the soft wood of *hala pepe* into idols (Lamb 1981).

Description

Hala pepe is a member of the agave family. It is related to *kī*, or *ti (Cordyline fruticosa)*, a plant introduced by Polynesians for thatch, food wrapping, raincoats, footwear, and other uses, and dracaenas *(Dracaena)*, a group of plants introduced much later by westerners for landscaping. It is an attractive plant and would make a nice landscape substitute for any of its introduced relatives. *Hala pepe* wood is soft but brittle and its branches can break in either strong winds or careless hands.

Fruits and Seeds

Hala pepe fruits resemble small cherries and are borne in clusters of twenty-five to perhaps two hundred on a projecting stalk. The fruits usually ripen during the summer and fall. The unripe fruits are hard, greenish, or khaki colored and wrapped in a thin husk of leaflike sepals. As they ripen, the fruits become soft and turn orange or red; often the husk erodes or is shed. The staggered ripening of individual fruits on a stalk can occur over four to eight weeks.

One or two white or yellow seeds form inside each fruit. Seeds removed from full-sized but unripe fruit are susceptible to fungi and often rot. In contrast, seeds removed from soft ripe fruit are more resistant to fungi and bacteria and show a high percentage of sprouting.

If you harvest a fruiting stalk with primarily unripe but full-sized fruit, treat it as you would a cut flower. (Also, before you remove it, make sure you have a home for all the young *hala pepe* that stalk potentially holds. Otherwise, just collect the ripe fruits.) Place the stalk's base, well below the fruits, in a jar or vase of shallow water. Every few days change the water and slice a few millimeters (¼ in) off the base to keep the water-absorbing tissues fresh. The *hala pepe* stalk will survive and its fruit will continue to ripen, providing viable seeds sometimes for nearly two months.

Germination and Seedling Growth

Using our standard method, *hala pepe* seeds begin to sprout in two to four weeks. Watch for a stout, whitish root to emerge from the seed. Over another week or two, the root grows a centimeter (⅜ in) or more before the shoot begins to take form. Initially, the shoot looks like a miniature lily.

If you keep your seeds outdoors, take the precaution of covering the seedbed trays with screening. Otherwise the seeds may be destroyed before sprouting by the larvae of an unidentified moth. This happened to a batch of sown seeds in Waimānalo, Oʻahu. More than a hundred seeds, under damp moss, were bored through and killed in just two weeks.

Pot the seedlings in a moist, well-drained potting mixture once the roots are 2–3 cm (1 in) long. The plants grow fairly rapidly; they seem to like a little peat moss near the surface of the soil. Any chlorosis (lightening or yellowing of the leaves) disappears quickly after giving the plant a small amount of general purpose fertilizer.

Repotting and Planting Out

Healthy *hala pepe* reach 20–30 cm (8–12 in) high in three or four months, at which time you should plant them out in well-drained soil. They thrive in full or partial sunlight; languish in deep shade. You can expect a growth rate of 30–50 cm (12–20 in) per year for healthy plants. For the first year or so after planting out, water your *hala pepe* well during dry spells. After two or three years, however, they will need almost no watering at all.

Young *hala pepe* sometimes develop a spotting disease on their leaves that can spread to the base of the plants and kill them. Too much fertilizer and overwatering are somehow related to this condition. Promptly cutting away the necrotic portion of the leaves with sharp, clean scissors can help prevent the spread of this disease. Some dense clay soils seem to promote root rot in cultivated *hala pepe*. This may be due to microbial contaminants in suburban lots, rather than soil type, as we have seen natural *hala pepe* groves on clay soils such as in the northern Koʻolau foothills of Oʻahu.

Scale insects occasionally infest *hala pepe*, particularly if the plant is stressed or already unhealthy. Remove the pests by hand if possible, rather than further stress the plant with an insecticide treatment. Larger organisms, especially root-chewing arthropods, will attack and sometimes kill young *hala pepe*. We have seen plants with stems up to a centimeter (⅜ in) thick girdled just at the soil line. Use a granulated insecticide such as Diazinon around the stems of young *hala pepe* if this problem is severe.

Hao
(Rauvolfia sandwicensis) Easy

Habitat

Hao occurs in mesic and dry forest, dry shrubland, and occasionally in wet forest and on lava flows. It is present on all the main islands except Kahoʻolawe (Wagner et al. 1990).

It is tempting to harvest these green *hao* fruits, but you would be wasting your time (and the fruits) because the seeds within the immature fruits of many Hawaiian plants will not sprout. Instead, try looking on the ground beneath a *hao* for viable seeds. The attractive leaves of *hao* contain a milky sap that seems to deter most leaf-eating insects.

Hawaiian Uses

Hao wood has been discovered in *heiau* (temples), where it may have had some religious uses (Lamb 1981).

Description

Hao is an attractive tree or shrub 3–10 m (10–33 ft) tall, with light green, glossy leaves clustered at the branch tips. The leaves have a prominent light-colored midvein. Also clustered at the branch tips are small white flowers. Two distinctive characteristics of *hao* are the tufts of small deciduous stipules (scalelike growths) at the leaf axils and the plant's milky sap. The roots of *hao* contain reserpine, an alkaloid used to treat high blood pressure and some mental diseases (Degener and Degener 1957).

Fruits and Seeds

The fruit is a bilobed drupe that turns from green and hard to purplish black and soft when ripe. Within are two (sometimes one) hard, flattened seeds about 1 cm (⅜ in) long with a rough outer surface. Ripening of *hao* fruits is quite variable from year to year; you may find them during late fall, winter, or even spring. When collecting seeds, collect only those from mature black fruits. It is our experience that seeds from immature fruits fail to germinate.

Germination and Seedling Growth

Plant *hao* seeds in the standard manner. Germination takes about four weeks (sometimes more) and the percentage of germination is usually high. Transplant the seedlings from the vermiculite bed to individual containers after they have two true leaves. Seedling growth is rapid—about 5 cm (2 in) per month. We have not encountered any seedling diseases or pests.

Repotting and Planting Out

Hao tolerates repotting well but is best placed in the ground after reaching 20–30 cm (8–12 in) in height. Plant in full sun to partial shade. New plants may need temporary support if winds are strong. *Hao* responds well to both dilute foliar and granular fertilizer. After outplanting, plants can grow as much as a meter per year. While *hao* has a wide tolerance for watering, anywhere from lightly once a day to heavily once a month, do not overwater. On rare occasions of extreme rains, we have seen the leaves on *hao* plants become flaccid and limp (an early sign of overwatering).

Hao has few pests, although occasionally a plant becomes infested with scale insects that are usually brought by ants. Eliminate the scale with an insecticidal soap and the ants with a granular insecticide such as Diazinon. Some butterfly and moth larvae also have a taste for *hao* leaves and stem tips. This is not a serious threat in our experience; simply remove the larvae by hand.

Hō'awa
(*Pittosporum confertiflorum, P. flocculosum, P. hawaiiense,* and *P. hosmeri*) Intermediate

Habitat

Ten native species of *hō'awa* inhabit the forests of the Hawaiian Islands. They range in size from shrubs to medium-sized trees. While most are adapted to mesic conditions, some species live near sea level while others grow at over 2,000 m (6,600 ft). *Pittosporum confertiflorum* has a wide habitat range, from dry to wet forest and even subalpine forest. It naturally occurs on O'ahu, Lāna'i, Maui, and Hawai'i. *P. flocculosum* is an O'ahu species growing in mesic to wet forest. *P. hawaiiense* and *P. hosmeri* both occur only on the Big Island. They live in mesic to wet forest and are very closely related. Where they occur together, they produce hybrids (Wagner et al. 1990). Some mesic adapted *hō'awa*, such as *P. hawaiiense*, do well in fairly dry settings provided they receive some watering during the drier parts of the year.

Hawaiian Uses

Hawaiians used *hō'awa* wood for the gunwales of canoes (Krauss 1993). They also treated skin sores with the crushed outer layer of *hō'awa* fruits (Bornhorst 1996).

Choose a *hōʻawa* appropriate for your location. This one is doing well at Queen Kapiʻolani Garden near the Honolulu Zoo. In days past, the now rare *ʻalalā*, our endemic Hawaiian crow, would feast on the hard, wrinkled seeds inside the ripe, walnutlike *hōʻawa* fruits.

Description

Hōʻawa is yet another example of a phenomenon that evolutionary biologists refer to as adaptive radiation (see *ʻākia, ʻakoko, koʻokoʻolau, kōlea, kōpiko, loulu,* and *pilo* for other good examples). According to evolutionary theory, the adaptive radiation of *hōʻawa* probably started long ago with a single ancestral seed, possibly from a Fijian plant (Wagner et al. 1990) hitching a ride with a bird to Hawaiʻi. That seed germinated and grew, and the founding tree reproduced to establish a viable population. Over time, seeds from this population were spread by other birds to other islands and other island environments. Isolated from the parent population, each new population changed over time to become a distinct species, either because of random events and mutations or by natural selection.

At the beginning of this century, Joseph Rock, affiliated with the College of Hawaii (now the University of Hawaiʻi at Mānoa) and an astute naturalist of the time, observed that nearly 80 percent of the fruit capsules of the *hōʻawa (Pittosporum hosmeri)* in Kona, Hawaiʻi, were eaten out by what was then the "very common" *ʻalalā,* or Hawaiian crow *(Corvus tropicus)* (Rock 1974). Today the *ʻalalā* is one of the most endangered native birds in Hawaiʻi, and the Kona *hōʻawa* has become exceedingly rare. Perhaps the crow was important to the seed dispersal of the *hōʻawa,* and now both species are nearly extinct.

Many *hōʻawa,* with leaves tufted near the branch tips, are very attractive. While the older foliage is usually dark green, new growth is often colored by a dense, light yellow to reddish brown hair. *Hōʻawa* flowers are white or yellow, tubular, and subtly scented; they occur in clusters along slender branches.

Fruits and Seeds

Fruits ripen at different seasons for the different species of *hō'awa*. For example, *Pittosporum confertiflorum* and *P. hosmeri* usually ripen during the winter months, while we have found *P. hawaiiense* fruiting in the summer. *Hō'awa* seeds develop in peculiar capsules that in most species look like walnuts. Typically, one or two capsules develop from each flower cluster. The green capsules gradually ripen, turning beige or brown, and then slowly open to reveal a hollow chamber packed with seeds. The interior of the capsule is often brightly colored—red or yellow—perhaps to attract birds, but there is no pulp in this strange fruit. The seeds are invariably dark brown or black and coated with a shiny, sticky, mucuslike material. The largest *hō'awa* seeds look like flattened raisins, but they have a hard consistency.

Germination and Seedling Growth

Given our standard bleach treatment and left to germinate in vermiculite beds under moss, *hō'awa* seeds may take many months to sprout. One batch of *Pittosporum flocculosum* seeds from O'ahu finally began to sprout after ten months. But after immersion of freshly harvested seeds from a ripe fruit of *P. hawaiiense* in lemon juice for twenty-four hours, followed by planting under moss as above, sprouting ensued in only one month. *P. flocculosum* may just take longer to germinate than *P. hawaiiense* or perhaps the acid in the juice (with a pH near 2) prompted sprouting by chemically scarifying the seeds. Is it possible that, in nature, these seeds sprout after exposure to the digestive juices in the guts of birds attracted by the colorful interior of open capsules?

Unlike the situation with some plants, collecting mature green fruits and allowing them to age in order to obtain good seed does not seem to work well with *hō'awa*. Often the capsules fail to open on their own and when finally cut open, the seeds appear shrunken and sometimes moldy—never healthy. We have never succeeded in germinating such seeds. Instead, it is best to collect only newly opened, mature fruits and process the seeds right away.

Hō'awa seedlings grow slowly. Aphids and red mites are attracted to young growth; they are best controlled by simply wiping the leaves with a damp cloth, paper towel, or small cotton swab.

Repotting and Planting Out

The *hō'awa* species with which we have worked all tolerate transplanting well. Growth often occurs in spurts and accelerates once branching begins in plants over 30 cm (1 ft) high. *P. hosmeri* seems to grow somewhat more slowly than the others, reaching a height of half a meter (20 in) after one year of growth in the ground. Red mites, aphids, mealybugs, and scale insects occasionally infest older plants. Control

ʻIliahi
(*Santalum ellipticum*, *S. freycinetianum*, and *S. paniculatum*) — Intermediate

Habitat

ʻIliahi grow over a wide range of terrain and climate. *Santalum ellipticum* is a coastal and dryland species on all the main islands except Kahoʻolawe; it is rare on Kauaʻi and Hawaiʻi. *S. freycinetianum* is common in the mesic forests of Kauaʻi, Oʻahu, Molokaʻi, and Maui but also occurs in the dry forests of Lānaʻi. The Lānaʻi variety, var. *lanaiense*, is federally listed as endangered. *S. paniculatum*, on Hawaiʻi, occurs on recently vegetated lava flows from 500 m (1,700 ft) all the way up through mesic and wet forest at 2,000 m (6,600 ft) elevation (Wagner et al. 1990).

Hawaiian Uses

Early Hawaiians had relatively little use for *ʻiliahi*. Occasionally they used the wood for musical instruments or for scenting finished *kapa* (Krauss 1993), but this changed after Western contact. The fragrant heartwood in various *Santalum* species had been valued in South and East Asia since antiquity. The discovery of new aromatic sandalwoods in Hawaiʻi around 1790 began the development of international commerce in the Islands.

The main sandalwood markets were in China, and cargoes of the newly valuable

This coastal *ʻiliahi* (*Santalum ellipticum*) in one of our gardens is proof that not all sandalwoods are difficult to grow. A young mountain *ʻiliahi* (*Santalum freycinetianum*) grows perched over a valley in the Waiʻanae Range, Oʻahu.

commodity, hacked out of the foothill and mountain forests, were carried for decades by westbound sailing ships out of ports from the Big Island to Kaua'i. The trade peaked in the 1820s as Hawaiian chiefs became sandalwood barons. They ordered forced-labor battalions of commoners into the mountains to cut and carry the fragrant wood. Western observers reported pharaonic spectacles—trains of thousands of people carrying sandalwood down to the waiting ships. In return, the chiefs got fancy trade goods from abroad through a mail-order business dominated by Americans; the commoners got nothing but grief and despair as whole villages were taken away from their subsistence labors, sometimes for weeks.

Perhaps surprisingly, these historically important trees survived and in some areas have made a modest comeback. On O'ahu, *S. freycinetianum* still grows on ridges behind Honolulu, above the North Shore, and in the Wai'anae Range. On the Big Island, one variety of *S. paniculatum* is common along Highway 11 near Kīlauea Caldera and the Volcano Golf Course, while another variety that resembles *S. freycinetianum* grows from Ka'ū to North Kona.

Description

The coastal variety (var. *littorale*) of *Santalum ellipticum* does not achieve tree stature but grows rapidly into a spreading shrub with blue green foliage. It naturally flowers in spring and summer with clusters of tiny cream-colored blossoms. Flowering often occurs year-round in cultivated plants. It is an attractive plant, quite suitable for a xeriscape garden. It is also the easiest of the Hawaiian sandalwoods to grow. The other Hawaiian sandalwoods are upright shrubs or trees from 1 to 2 m to over 15 m (50 ft) in height, with rugged bark, abundant foliage, and clustered red *(S. freycinetianum)* or yellow brown *(S. paniculatum)* flowers.

Fruits and Seeds

Ripe *'iliahi* fruits resemble small purplish black olives. While individual plants usually have ripe fruits for only several weeks at a time, the fruiting season for most species lasts from spring through fall. Some plants develop more than one crop of fruits during the year. The seed is a single usually spherical to ovoid pit, off white to khaki in color with a rough-textured surface. Do not bother attempting to sprout seeds from green fruits; they seldom produce strong seedlings.

Germination and Seedling Growth

Prior to our research, growers reported waiting months for *'iliahi* seeds to sprout. Sometimes seeds would not sprout at all. Now, after several experiments (with *Santalum ellipticum* and *S. freycinetianum;* we have not yet tested *S. paniculatum*), we can state with some confidence that using a pretreatment of gibberellic acid (a plant hormone) results in the rapid germination of viable seed. You can purchase this hormone from a biological supply company such as Carolina Biological Supply, Inc. Unfortunately, the small amounts needed require a sensitive scale for weighing, but we hope resourceful growers will overcome this inconvenience. Without the hormone treatment, gardeners using our general methods will still be able to sprout sandalwood seeds, but germination will take much longer.

Begin by extracting the seeds from ripe fruits and cleaning them by hand. Let the seeds air-dry for approximately one week. Then, using forceps or medium sandpaper, remove a small portion of the seedcoat at the pointed end (apex) of the seed so that the embryo inside is visible; do not damage the embryo. Soak the seeds in a shallow container of 0.05 percent gibberellic acid for five days, changing the solution daily. Afterward, remove the seeds from the gibberellic acid solution and dust them with a 1:1 mixture of powdered sulfur and Captan. This will inhibit fungus from infecting the seeds. Sow the seeds in a covered tray on new, moist vermiculite.

The seeds begin sprouting in about one week and continue to germinate for another two or three weeks. Watch for a crack to develop in the thick wall of the seed. The crack's interior, the embryo, looks white against the dark seed coat. As the crack widens, the root emerges from the embryo's apex. After good root development (an additional one to two weeks), transfer the seedlings to individual pots containing a 1:1 mix of fine cinder (not black sand) and vermiculite. (Using a peat-based soil caused some seed and seedling death by fungal infection.) Germination using this method is over 90 percent successful. Even seeds as old as seven months responded to this treatment and sprouted at a high percentage.

Healthy sprouts quickly send up a robust shoot that is largely resistant or unattractive to insects, except for those that gnaw stems at ground level. The usual suspects are cockroaches, sowbugs, crickets, and various cutworms. Some snails and slugs may also attack newly sprouted *'iliahi*. Such pests can fatally girdle a seedling in one or two nights. This problem is rare and easily controlled in seedling containers, but it can be a major headache in new outside plantings.

Repotting and Planting Out

Unlike most Hawaiian plants, *'iliahi* should be planted in the ground as young seedlings. They rarely develop beyond this stage in pots. We have succeeded with *'iliahi* planted out from two to six months of sprouting. You can plant the seedlings directly from their vermiculite sprouting trays or later, from the 1:1 fine cinder:vermiculite mix.

In planting out *'iliahi*, there is an exception to our general guideline on the size of the hole to dig. It should be just large enough to comfortably accommodate the plant's root structure. This is because of the seedling's probable need to develop a symbiotic association with the roots of other nearby plants (see discussion below). If you have to transplant your *'iliahi* after several months, try to avoid excessive root damage and exposure. Several *Santalum paniculatum* tolerated replanting from one landscape site to another three months after the first planting.

S. ellipticum and *S. paniculatum* like sunny conditions; *S. freycinetianum* may do well in partial shade. For the first week or two after planting out, shelter the young plants from full exposure to the sun, gradually tapering off the artificial shade. Because the plants are small with shallow roots, you must keep them well watered for at least several months.

Typically, *'iliahi* do not grow for at least several weeks to months after planting into the ground. Sandalwood development appears unique in that seedlings seem to have a certain "waiting period" in which, having exhausted the energy stored in the seed, they exist in stasis until establishing the necessary root contact with other species. We suspect this phenomenon occurs in all Hawaiian *Santalum* species. The waiting period in the three species of *'iliahi* we have grown seems to be limited to approximately one year, after which the plants, still just 10–15 cm (4–6 in) tall, gradually decline. Their leaves yellow and fall off, and they die over the following few weeks or months.

You will know if you have been successful in establishing your *'iliahi* when a dramatic burgeoning begins. Often the new growth (in *S. ellipticum* and *S. paniculatum*) shows a change in color—it may be initially red or pink, then bluish green. New foliage is added rapidly, indicating the plant has taken hold.

Sandalwoods are believed to be root parasites on other species. The sandalwood roots develop tiny hairlike extensions that capture some essential nutrients from other roots they encounter in the soil. Successful establishment of mountain sandalwoods may require native host plants, such as *koa* trees (reported to be the most common host) or other native species, to be present nearby.

The coastal *'iliahi*, *S. ellipticum*, of course, does not occur with *koa*, but may also require a physiological connection to the roots of other plants. In Waimānalo, several of our *S. ellipticum* have grown large over five years in a planting with *'ākia* (*Wikstroemia sandwicensis*) and the Nihoa *loulu*, *Pritchardia remota*. An earlier attempt to plant out *'iliahi* in this patch of ground, before establishment of the *'ākia* and *loulu*, failed. On the other side of O'ahu, our healthy two-year-old *S. ellipticum* plants are growing with *Chenopodium oahuense*, *Chamaesyce hypericifolia* (an alien weed), or only introduced grasses growing nearby. Considering these successes, it seems that *S. ellipticum*, if indeed it does need a host plant, is not very specific in its requirements.

S. ellipticum thrives in sandy soil near the beach as well as the red clays of Leeward O'ahu. Two of four shrubs in Waimānalo flowered in the second year, but they did not set fruit until the third. On the leeward side, three of four plants flowered during their first year and one of four set fruit during that first year. In 1996, coauthor Culliney established *S. paniculatum* seedlings in a suburban yard near Volcano, Hawai'i. After one year these mountain sandalwoods, planted within 10 m (33 ft) of a robust five-year-old *koa* tree and mature *'ōhi'a lehua*, averaged 30 cm (1 ft) tall. After two years they are ½–1 m (1½–3 ft) tall, growing rapidly, and starting to branch.

We have tried planting out *S. freycinetianum* only on a couple of occasions. Unfortunately, these attempts have failed.

If necessary, you can prevent insect damage to the stems of young plants near ground level by using the nontoxic sticky resin (trade name Tangletrap or Tanglefoot) available in garden shops. Apply a thin layer all around the lowest 2–3 cm (1 in) or so of the stem. If particles of soil and organic debris get caught in it, so much the better. The goop and junk in it will gradually coalesce into a kind of armor coat that protects the stem for weeks. Check on it occasionally, however; you may need to reapply the resin once or twice before the plant develops some bark at its base to take over the protection.

'Iliahi foliage resists most insect attack. Mild infestations of whitefly or scale insects sometimes occur. Occasionally, a small gray weevil will eat the younger leaves. This can usually be ignored; for treatment use insecticidal soap or acephate.

'Ilie'e
(Plumbago zeylanica) — Easy

Habitat

'Ilie'e occurs in dry lowland areas including dunes, shrubland, and forest. It is found on all the main islands (Wagner et al. 1990).

Hawaiian Uses

Hawaiians used the sap of *'ilie'e* to blacken tattoos. In Asia and Europe the plant is used medicinally; there is some evidence that Hawaiians also used it for medicine (Neal 1965).

Description

Hawai'i's remaining dryland forests seem to have few native understory plants. Whether this is their natural state or the result of centuries of herbivory by introduced animals such as cattle, goats, and pigs—as well as competition with alien grasses and other weeds—we may never know for certain. *'Ilie'e* is one of the few native understory plants we see today. Perhaps the fact that it is poisonous to eat (Neal 1965) is responsible for its persistence.

Technically, with its many long stems originating from a single base, *'ilie'e* is considered a sprawling shrub. But because these stems can be over 2 m (6½ ft) long and grow close to the ground, *'ilie'e* can easily be mistaken for a vine. Planted closely together and pruned to promote branching, *'ilie'e* makes an attractive and effective groundcover around the base of trees or shrubs. Its small, white, star-shaped flowers

Each sticky green fruit of *ilie'e* contains one small seed. More a sprawling shrub than a climbing vine, *'ilie'e* makes an excellent native replacement for the ubiquitous and alien *Wedelia* groundcover.

will bloom throughout the year if the plant is well watered. Under very dry conditions, such as those in a native dry forest in summer (or an unattended garden), *'ilie'e* will lose its leaves and its stems will die back to the plant's base. Yet when the winter rains arrive, expect *'ilie'e* to burst back to life and quickly grow new stems and leaves to cover the forest floor.

Here is an interesting observation for land managers and other conservationists: In the understory of both alien and native dryland forests, we have seen *'ilie'e* and no grass or grass and no *'ilie'e*. Does *'ilie'e* competitively exclude grasses and vice versa?

Fruits and Seeds

'Ilie'e fruits are about 1 cm (⅜ in) long, green or brown when ripe, and occur in clusters at the tip of a stem, often near still-blooming flowers. They are very sticky, an adaptation that aids their dispersal by birds or pant legs. Each fruit contains one small (3–4 mm; ⅛ inch), hard, dark brown seed. If, when dissecting fruits for the seeds, you find a seed that is still yellow or green, discard it; the seed is not mature.

Germination and Seedling Growth

Plant *'ilie'e* seeds in the standard manner, except that we have never sterilized the seeds with household bleach. The seeds take three or four weeks to germinate. Transfer the seedlings to individual containers about one month after sprouting or when they have two or three sets of true leaves. Begin watering the seedlings with a dilute liquid fertilizer (or use time-release fertilizer pellets) to speed their growth. Seedlings sometimes develop a sticky white powder on their leaves and stems. This is a secretion from the plant and is easily washed off with water.

You can also propagate *'ilie'e* from stem cuttings. Use cuttings of semihard green stems following the procedure in the previous section. The majority of the cuttings will develop roots in less than a month. Wait for good stem growth before repotting the new plant.

Repotting and Planting Out

'Ilie'e grows well in pots and can make an attractive native Hawaiian hanging basket. Fertilize and prune it often to keep the shape you desire. If you want to plant out your *'ilie'e*, pick a location with partial shade. It requires little or no supplemental watering after it is established (in two to three months). *'Ilie'e* can also be planted out in full sun but will require more water to remain lush and flowering. In the ground, the plant grows quickly; a single stem can grow as much as a meter (3 ft) in one to two months. We have not encountered any pest or disease problems with *'ilie'e*.

Kauila
(Alphitonia ponderosa)   Intermediate

Habitat

Kauila occurs on all the main islands except Kahoʻolawe and Niʻihau. It grows in dry to mesic forests from around 250 m to 1,200 m (825–4,000 ft). Only Kauaʻi has large numbers of this otherwise rare tree (Wagner et al. 1990).

Hawaiian Uses

The Hawaiians considered the dense, hard wood of *kauila* an important material for *hale* construction (e.g., house beams), heavy tools and weapons (e.g., *kapa* mallets, *ʻōʻō*, and spears), and for game pieces and musical instruments. They also made a bluish dye from the leaves and bark (Krauss 1993).

Description

There are two endemic species named *kauila* by the Hawaiians. Both are members of the buckthorn family (Rhamnaceae). *Alphitonia ponderosa* has proved a somewhat better survivor than its relative *Colubrina oppositifolia*. The most reliable place to encounter *Alphitonia* in the wild is on the leeward slopes of Kauaʻi. Here, significant populations still grow in the western parts of Kokeʻe State Park. The Big Island's South Kona forests also contain some mature and young *Alphitonia* trees.

    Alphitonia grows to a maximum height of 25 m (80 ft), but it is often considerably smaller than this (Wagner et al. 1990). The foliage is attractive, featuring strongly veined, sharply pointed leaves. The bark is thick and rugged, with little tendency to peel. *Alphitonia* trees probably grow slowly in the Hawaiian dry forests but, under cultivation, some of our young trees have grown fairly rapidly in well-drained soil even close to sea level.

If pollinated, each of the small flowers of *kauila (Alphitonia ponderosa)* will develop into a hard acornlike fruit containing two or three seeds. There is little doubt that Hawaiians could tell the difference between the two *kauila,* even when the trees were very small. They gave the two a single name probably to emphasize their similarity, the hard durable wood of both being used for tools and weapons.

Fruits and Seeds

Alphitonia fruits are approximately fingertip size. When ripe, normally between May and September, they consist of a hard, brown to black outer husk with a texture like masonite that covers two or three thinner-walled inner capsules containing the seeds. The inner capsules are actually stronger than the outer husk. To open the dry, mature fruits with the least trauma to both seeds and your fingers, use pliers to collapse the thick outer husk. Remove the inner capsules and, if needed, shave away any remaining husk material with a penknife. Place the capsules in a dry place or, if you are in a hurry, in sunlight or under a warm lamp. The capsules will open by themselves within an hour or two in response to warmth and drying. You can then easily extract the seeds.

The seeds are about 6 mm (¼ in) long, dark brown to black, and somewhat flattened. They have the outline of a truncated oval with a slightly ribbed surface. A light brown parchment covers the seeds when they are in place in the inner protective capsule. This parchment is easily removed, and seeds sprout faster without it. Scratching the seed coat with a penknife or sandpaper also speeds germination.

Germination and Seedling Growth

Alphitonia seeds swell to perhaps twice their dry size as they take up water over the first one or two weeks but, commonly, require from two weeks to two months to sprout roots that emerge from the truncated ends. Some batches of seed seem to be highly susceptible to fungal or bacterial growths that kill the embryos, while others are resistant. Once the seeds begin to sprout, they become more resistant to such diseases as long as they are well drained. The roots are thin and white and grow rapidly. You can transfer sprouts to individual pots as soon as you see roots; use a potting mixture of about half small cinder and half vermiculite; keep well watered. Sulfur in the potting mix will protect the young seedlings against many diseases (see previous section). Shoots and seed leaves appear a week or two after the emergence of the root. You can also safely leave the sprouts in the vermiculite seedbed tray and transplant after their early leaves have developed. Young seedlings are sometimes attacked by a fungal stem rot that also plagues the other *kauila* species. See the next entry on *Colubrina* for a suggested treatment.

Repotting and Planting Out

Young *Alphitonia* are susceptible to drying of the soil. Of a half dozen young plants transplanted into the yard of a vacation home near the Volcano Golf Course—a mesic location on the Big Island at 1,200 m (4,000 ft)—all but one died during periods of drought when the plants were not watered. The lone survivor, now five years old, is still less than a meter high. By contrast, well-watered potted specimens at sea level reach an equivalent height in about one year. However, potted trees, sensitive to high soil temperature, stop growing during the hottest summer months.

The black twig-borer will attack young and older *Alphitonia*. An attack on the main stem of a young tree will kill it. When stressed by disease or drought, *Alphitonia* seem less able to ward off an attack by this lethal pest. Whiteflies and spider mites occasionally infest the leaves of *Alphitonia*. If the population becomes large they cause the premature loss of older leaves; they may also stress the plant, making it more susceptible to attack by the black twig-borer. Eliminate the foliage pests either by hand with a moist towel or by spraying with insecticidal soap. See the previous section for information on deterrents and treatments for the black twig-borer.

Kauila
(*Colubrina oppositifolia*)   Intermediate

Habitat

One of Hawai'i's rarest native trees, significant numbers of *Colubrina oppositifolia* occur only on the dry leeward slopes of the Big Island from about 200 m (660 ft) to perhaps 1,000 m (3,300 ft). Elsewhere this beautiful tree is extinct or nearly so (e.g., O'ahu). It is federally listed as endangered.

Hawaiian Uses

Hawaiians used the exceedingly hard, heavy wood of *Colubrina* interchangeably with *Alphitonia* for heavy tools such as *kapa* mallets, poles for *kāhili* (a tall banner used in war and ceremony), weapons, and so on. *Kauila* splinters were reportedly fashioned into hairpins (Lamb 1981).

The bark of the endangered *Colubrina oppositifolia* looks composed of many interlocking puzzle pieces. This young *kauila* is afforded at least some protection from fire within the tiny grass-free State enclosure off Māmalahoa Highway in North Kona, Hawaiʻi.

Description

For mainlanders, this species resembles an apple tree with its dappled gray bark arranged in patches that seem to form a natural jigsaw puzzle. The leaves are reddish and glossy when young and duller when mature, with peculiar swellings on their lower surfaces.

Nearly all the remaining wild *Colubrina oppositifolia* grow on ranch lands in North Kona, Hawaiʻi. Tragically, the increasingly common brush fires in this region have killed many of these extremely rare trees, together with other endangered native plants. The fires, combined with the deleterious effects of cattle and alien grasses—particularly the fire-adapted fountain grass *(Pennisetum setaceum)*—are rapidly extinguishing this one-of-a-kind Hawaiian dry forest ecosystem. The only other place we have seen wild *kauila (Colubrina)* is in Oʻahu's Waiʻanae Range, where a single grove of trees is near extinction.

Fruits and Seeds

Colubrina fruits mature from late spring through early fall. The fruits look like those of *Alphitonia,* but are slightly smaller. Greenish brown, mature (full-size) fruits have often yielded healthier seeds than older brown fruits. But let these fruits ripen for a few weeks in a dry, clean place, protected from insects, before extracting the seeds. (Do not collect entirely green fruits; they are too young.) *Colubrina* seeds, like those of *Alphitonia,* are enclosed within tough capsules surrounded by a woody-fibrous husk. But the thin parchment that covers the seeds of *Alphitonia* is absent in *Colubrina.* The seeds are about 6 mm (¼ in) long, smooth, and dark brown to black when mature. Upon sprouting, the root emerges from the more sharply tapered end of the seed.

Not all *Colubrina* trees produce good seed; often we have collected seeds that lack an embryo. You can test to see if a seed contains an embryo by squeezing it between your fingers. Those without an embryo flatten with modest pressure and can be discarded. Yet even seeds that pass this test are often susceptible to fungal or bacterial rot that develops despite our standard bleach treatment. Possibly the rot-causing spores have colonized minute openings in the seeds where they elude the disinfectant. These seeds rot after they swell with water, and soon they are covered with bacterial and fungal growths. Dissection of the rotten seed reveals that it is primarily the white tissue around the cotyledons (seed leaves) and the hypocotyl (the stem below the cotyledons) or the embryonic root that are rotten. These parts turn mushy and disintegrate, while the green seed leaves and embryonic stem tissue remain intact.

Germination and Seedling Growth

Most healthy seeds, without scarifying, swell noticeably within twenty-four to forty-eight hours after soaking in water and sowing in clean, moist vermiculite and sprout within two to four weeks. A thin white root emerges and grows rapidly (several millimeters per day). Expect anywhere from 25 to 75 percent germination, based on our experience. You can plant sprouts immediately in clean, moist potting mixture—equal portions of small cinder and vermiculite—or you can wait until the young seedlings develop one or more sets of true leaves. If you lose many seeds or seedlings to fungal diseases, try watering from the bottom only. If you transplant sprouts, it is essential not to let the subsoil dry out. The shoot and seed leaves of healthy sprouts will develop within a few days.

One-month survival of seedlings from healthy seeds is excellent (80 to 100 percent). But like *Alphitonia, Colubrina* sometimes develops a damping-off degenerative stem disease after one or two months. Watch for small white patches on the stems that, if untreated, spread and turn dark after several days. The final, fatal stage is a shriveling of the damaged portion of the stem. You can successfully halt this disease in its early stages with powered sulfur applied directly to the afflicted stem with a toothpick. Several applications may be necessary, spaced over a couple of weeks; healthy barklike tissue forms over and heals the treated lesions. Treating the seedling near the soil line seems especially important, but stem infections may occur anywhere up to the seed leaves. You can also routinely treat healthy seedlings, stroking sulfur powder on stems with a toothpick about once a week, to prevent the stem rot problem.

Seedlings and older plants sometimes contract a disease that deforms their leaves. As the leaves develop they begin to wrinkle and shrivel, more or less severely. Young plants also begin to branch profusely. Left untreated, the problem may worsen so that much of the foliage becomes grossly distorted. The most effective treatment seems to be to remove the distorted foliage. Most plants begin producing normal leaves after one or two rounds of selective pruning.

Repotting and Planting Out

Kauila tolerates transplanting well. With modest fertilization and daily sunlight, six-month-old *Colubrina oppositifolia* will be 20–30 cm (8–12 in) tall. You should plant them out at this size. After you see new growth, resume fertilization every two or three months with any all-purpose fertilizer. Surprisingly, *kauila* begin flowering and fruiting at a very young age. Some of our plants, at less than a meter high and only two years old, have already flowered and fruited. After five years, some faster growing *kauila* can be nearly 2 m (6½ ft) tall. *Kauila* appear to thrive in a range of soil types and elevations, from heavy clay soils near sea level to more typical soils and elevations such as the ʻaʻā lava fields at 600 m (2,000 ft) in the small State of Hawaiʻi endemic plant sanctuary near Puʻuwaʻawaʻa, Hawaiʻi.

Few insect pests have bothered our plants. Occasionally a small gray weevil will eat the young leaves, but it often disappears without treatment. We have seen the black twig-borer attack plants in the wild and in moist shade houses, but none of the *kauila* we have planted out in urban settings have been attacked by this pest. Perhaps only stressed *kauila* are susceptible to attack by this borer.

Kāwelu
(Eragrostis variablis) Easy

Habitat

Kāwelu occurs in coastal dunes and grasslands, open sites in dryland forest, and on exposed cliffs up to 1,100 m (3,600 ft) throughout the Hawaiian Islands (Wagner et al. 1990).

Hawaiian Uses

Hawaiians would sometimes use *kāwelu* as an alternative to *pili* grass *(Heteropogon contortus)* for thatching (Krauss 1993).

Need a green fountain for your rock garden or a groundcover for a barren hillside? Try the native perennial bunch grass, *kāwelu*.

Description

Many state, county, and private gardens in Hawai'i are beginning to include more and more native plants in their collections. However, the natives seen are mostly dicots and the more showy monocots such as *hala pepe*. Almost no one in Hawai'i seems to be growing native grasses. *Kāwelu*, while not showy, is an attractive perennial bunch grass (more commonly referred to as a lovegrass) that grows to about a meter (3 ft) in height. A single *kāwelu* looks like a small, light green fountain (see photo). Best of all, *kāwelu* is easy to grow.

Fruits and Seeds

Kāwelu fruits are called caryopses or, sometimes less precisely, grain. Each caryopsis contains one seed. In *kāwelu*, hundreds of small caryopses, about 2 mm (1/16 in) long, are clustered on one or more spikes that typically grow taller than the grass' leaves. When ripe the caryopses are light brown. (The spike has usually turned brown by this time also.) You can find plants with ripe caryopses throughout most of the year. If possible, collect caryopses from more than one plant to maintain genetic diversity in your garden. Also, since each spike has hundreds of caryopses, resist taking the entire spike. Instead, cut off part of the spike or collect the numerous caryopses by hand, leaving some on the spike.

Germination and Seedling Growth

Plant *kāwelu* caryopses in the standard manner, but because of the small size of the seedlings, do not place a green moss covering over the seedbed. The seeds will begin germinating in one to two weeks. We have noticed that some of the seedlings grow more vigorously than others. Separate and transplant the seedlings to individual containers when they reach about 5 cm (2 in) in height and have four to six leaves.

Repotting and Planting Out

Do not let your *kāwelu* become pot-bound. Pot-bound plants become spindly and rapidly exhaust the water supply in the pot. To date, we have planted out only four *kāwelu* in one location. This site is in full sun and watered regularly. Two of the plants were planted side by side and within a year one had completely dominated the other (i.e., grown much larger). The dominant grass, after two years, flowered for the first time in winter. It takes about two years for *kāwelu* to grow half a meter (1½ ft) tall and a meter (3 ft) in diameter. This grass may be short-lived; recently a previously healthy three-year-old *kāwelu* died quickly with no signs of disease or pests.

We have not seen any insect pests other than the occasional grasshopper on our four *kāwelu*. We do not know what impact Chinese rose beetles would have on *kāwelu* because the planting site is partially lighted at night. Land snails and slugs also show some interest in young plants; older plants appear resistant to these molluscs.

Koa
(Acacia koa) — Intermediate

Habitat

Koa is commonly a tree of mesic forests but is also found in drier zones from near sea level (60 m; 200 ft) to slightly over 2,000 m (6,600 ft). It is still common on all the main islands except Kahoʻolawe and Niʻihau.

Hawaiian Uses

Hawaiians prized *koa* for construction, especially the hulls of oceangoing canoes, but also for paddles, spears, the shafts of *kāhili* (a tall standard used in war and ceremony), containers (but never for food such as poi because the wood imparted a bitter taste), and surfboards (the shorter variety; *koa* replaced *wiliwili* for long boards after European contact). The Hawaiians also made a red dye from *koa* bark (Abbott 1992; Krauss 1993).

One of the fastest growing and potentially the most majestic of Hawai'i's native trees is the *koa*, whose stature often lives up to its Hawaiian namesake in translation, the warrior. Hawaiian *koa* changes its leaves as it gets older from a feathery compound leaf to a sickle-shaped phyllode—not a real leaf but a flattened petiole. This was just one of the ways Hawaiians recognized their native *koa* from the alien *koa haole (Leucaena leucocephala)* introduced after Western contact.

Description

Beautiful and majestic, *koa* can grow to the largest size of any native Hawaiian tree (35 m; 115 ft), but there are many "types" that do not grow nearly as tall. *Koa* is an acacia, a tree legume, distantly related to garden beans and peas. Unlike its smaller, spindly relative, the alien weed tree known as *koa haole (Leucaena leucocephala)*, *koa* foliage changes during the tree's lifetime. While the young leaves of *koa* have a feathery, compound structure similar to *koa haole*, the mature foliage consists of sickle-shaped phyllodes—not true leaves at all, but flattened and elongated petioles (the leaf stems) of the compound leaves. As a *koa* tree matures, it produces fewer compound leaves and more phyllodes.

The most prized *koa* in ancient Hawai'i grew in steep-sided gulches where, in competition with other trees to reach the sunlight, the trunks grew straight and tall for many meters. These were the prime trees for canoe building. The trees were probably always rare, but by the time of European contact, Hawaiians were traveling many miles inland and thousands of feet up into the mountains to find them (Alexander 1934). Ethnobotanists such as Abbott (1992) have suggested that because these largest trees were more common on the highest islands of Hawai'i and Maui, they were in part responsible for the success of Hawai'i's and Maui's warring chiefs such as Kamehameha and Kahekili.

Felling and hollowing out a *koa* tree for a canoe included a substantial amount of ritual. First, a master canoe builder *(kahuna kālai wa'a)*, with permission from his chief, would find a suitable tree and give prayers and offerings to Kū. Carpenters wielding stone adzes would then fell the large tree and further prayers and offerings would be made. After the tree was felled but before it was shaped, the *kahuna kālai waʻa* kept a watchful eye for the *ʻelepaio,* a native forest bird believed to represent Lea, the wife of Kū. If the *kahuna kālai waʻa* saw this small bird alight and peck at the tree, he would declare the log unsuitable and it would be abandoned. Because the *ʻelepaio* is an insect-eating bird, its appearance on a downed *koa* was probably an indication that the tree's wood was infested with insects and would make a poor canoe. If there were no *ʻelepaio,* the carpenters would set to work shaping and hollowing out the log; this often took many days. Finally, the men would drag the roughly hewn hull many miles down the long slopes and valleys to a seaside *hale waʻa* (canoe house).

Western colonists of Hawaiʻi also learned to appreciate the woodworking qualities and great beauty of *koa.* It was and is referred to as "Hawaiian mahogany," even though it is not at all closely related to mahogany. Today local artists and woodworkers produce a variety of *koa* products, including containers, fine furniture, and picture frames. *Koa* lumber even has a valued reputation abroad.

Koa trees have been cut down in massive numbers on all the islands where it grows, and severe ecological damage may be the consequence of continued logging in native forests. Because *koa* is so vital a species—a kind of keystone in its ecosystem—nurturing the soil by nitrogen fixation, providing shade for the seedlings of other plants, and establishing a canopy habitat for native insects and birds, its wholesale removal triggers the collapse of a unique native biological community that takes many decades or even centuries to build. A reasonable approach to developing and exploiting a growing market for *koa* wood and products would be to plant it on the newly bare sugarcane lands and eroded ranches. There are encouraging signs that this idea is gaining ground, although in one much ballyhooed forestry project, the landowner cleared virgin old-growth native *koa* forest with its unique holistic natural heritage, then reseeded the area as a *koa* tree farm.

Fruits and Seeds

Depending on the elevation, *koa* produce fruits from fall through spring. The fruits are flattened pods 10–20 cm (4–8 in) long and about 1–2 cm (⅜–¾ in) wide. Mature pods are dark brown and brittle. You can easily split them lengthwise to harvest the brown to black seeds. Often, older pods have already split open by themselves; search the ground for the fallen seeds. Check carefully for insect damage when collecting seeds. Healthy *koa* seeds resist rot and you can store them for several years. For example, we collected a batch of seeds still in their pods and stored them in a sealed plastic bag. Three years later, 95 percent of these seeds sprouted and grew into healthy seedlings.

Germination and Seedling Growth

Koa seeds sprout readily after scarifying (rub with sandpaper or scratch lightly with a knife blade) and soaking in shallow water. Overnight the seeds swell up to two or three times their original bulk as they take on water. Roots emerge within a few days to a week. Shoots appear a few days after roots emerge. As with species such as *'āwikiwiki* and *wiliwili,* you can plant *koa* seeds immediately into individual pots containing new potting mix after scarifying and soaking.

Koa seedlings grow rapidly given adequate water and light. If you start seeds indoors, place the plants outside early to avoid spindly growth. In order to hasten their growth, begin to lightly fertilize your seedlings two or three weeks after they have two or more true leaves. Aphids and whiteflies sometimes infest seedlings. Eliminate these by hand with a damp towel or by spraying with an insecticidal soap. *Koa* seedlings are also attacked by a seedborne vascular wilt fungus *(Fusarium oxysporum* f. sp. *koae)* that kills them (Anderson and Gardner 1998). This fungus also kills adult *koa* (see below). Seedlings afflicted with this disease develop curled leaflets and leaves that tend to yellow, die, and fall off prematurely. Currently we know of no effective treatment for this disease and recommend that you isolate and discard any infected seedlings you encounter. Pretreating *koa* seeds with household bleach may help to prevent fungal infection.

Repotting and Planting Out

Koa trees achieve their explosive growth by virtue of a symbiosis with specialized bacteria that live associated with the roots (Rosa 1994). These bacteria absorb nitrogen gas from the air in the surface layers of the soil and convert the inert gas into high-nitrogen fertilizer that eventually surrounds the tree's entire feeding root system. This symbiosis is also common in a variety of related plants, among which are food crops such as peas and beans. In the wild, *koa* seedlings acquire the bacteria naturally as their roots begin to grow into the forest soil. If your *koa* seedlings grow slowly or have pale foliage after planting them out in your yard or landscape, it is possible they did not pick up the symbiotic microorganisms. To be certain that home-grown *koa* seedlings are inoculated with these beneficial bacteria, you can simply use a generous pinch of topsoil from a wild *koa* grove (common along many island roadways) in your potting mix. Another technique is to stir some *koa* soil into the water you use to water your seedlings. Some local nurseries now stock the *koa* inoculant and you can give your young trees a shot of this "medicine."

In mesic settings, several hundred meters or higher above sea level, *koa* planted out at less than a meter in height reach 3–5 m (10–17 ft) in as little as two years. Some of our five-year-old trees are now 10 m (33 ft) tall with trunks 30 cm (1 ft) thick. At lower and dryer elevations (100 m or less), growth has not been as rapid; after two years, these trees are only 2–3 m (6½–10 ft) tall. We should note, however, that the seeds we used for the above comparison came from different sources.

In parts of Hawai'i, below about 500 m (1,650 ft) elevation, a variety of pests (including, in our experience, Chinese rose beetles, the small gray weevil, mealybugs, whiteflies, and a stem-boring grub) attack *koa*. While larger trees may only be disfigured, attacks on roots, stems, and foliage are sometimes so severe that young trees are killed. Please refer to the previous section for treatments against these pests.

None of the *koa* (except for seedlings) we have cultivated have ever been attacked or killed by either the black twig-borer or the vascular wilt fungus mentioned earlier, but we have seen cultivated *koa* elsewhere suffer from both. Fortunately, research is now underway to find varieties of *koa* resistant to black twig-borer attack. Until then, we recommend you keep a vigilant eye for any signs of damage by this insect. If you detect and deal with the damage early, you can greatly minimize the threat of serious disfigurement or death. Certainly if you believe you have a *koa* that shows some resistance to the borer, do not keep it a secret. Your tree may be just the variety we are all looking for.

Unfortunately, the prognosis for *koa* infected by the vascular wilt fungus is poor. The common sign of infection is the rapid death and loss of leaves from the tree. Often this will happen initially to only one branch or one half of the tree, but soon the entire tree, or most of it, is leafless and dead. Sometimes stands of *koa* in the wild will die from this disease, surrounded by healthy *koa*. This suggests that either some *koa* are naturally (i.e., genetically) resistant to the fungus or that some other factor such as soil type or drought stress is responsible for the varied response. Again, research with which everyone can help offers the best hope of cultivating *koa* in areas where this fungus or its spores are present.

Koai'a
(Acacia koaia)   Intermediate

Habitat

Reportedly, *koai'a* occurs in dry, open areas on Moloka'i, Lāna'i, Maui, and Hawai'i (Wagner et al. 1990). We have seen only the Hawai'i population, found in west Kohala. It is our understanding that the Kohala population is the largest wild population of this threatened tree.

Hawaiian Uses

Hawaiians used the hard wood of *koai'a* for tools, fishhooks, *'ūkēkē* (a musical bow) (Krauss 1993), spears, and canoe parts such as paddles (Lamb 1981). Today woodworkers use the dead wood for gun stocks, knife handles, bowls, and artwork (Hancock, pers. comm.).

Description

For gardeners who do not have the space or the climate for a *koa* in their backyard (most *koa* do best under mesic conditions), a *koai'a* is a nice alternative. In contrast to *koa*, which can reach 35 m (over 100 ft) in height, *koai'a* is a small tree, usually less

The beautiful round flowers of *koaiʻa,* when pollinated, will develop into long thin pods. Through the earnest and unselfish efforts of Judy and Will Hancock of Kohala, Hawaiʻi, a double row of young *koaiʻa* trees extends as far as the eye can see.

than 5 m (16 ft) tall. Its wood is significantly harder than that of *koa* and structurally quite different (R. Hobdy, pers. comm. in Wagner et al. 1990). The phyllodes, the mature "leaves," are often but not always shorter and straighter than those of *koa.* The seed pods of *koaiʻa* are narrower that those of *koa,* with the long axis of the seeds arranged parallel to the long axis of the pod; in *koa* the long axis of the seeds is arranged parallel to the short axis of the pod. Wagner et al. (1990) consider *koaiʻa* a subspecies of *Acacia koa* rather than a separate species because of the discovery of an intermediate form found in a mesic forest in northern Kauaʻi. We consider this overly conservative taxonomy and have retained (above) the distinct species name.

The efforts of Judy and Will Hancock of Kohala, Hawaiʻi, are a dramatic demonstration of the power that private individuals can have in rescuing our endangered flora. In the past ten years, the Hancocks have planted over 2,300 *koaiʻa* on their ranchland. Protected from cows and other herbivores by electric fences, the seedlings have grown into vigorous trees 2–3 m (6½–10 ft) in height, with an abundance of flowers (see photo).

Fruits and Seeds

The fruits (pods) of *koaiʻa* look similar to those of *koa* except for the above-mentioned seed orientation. When ripe, the pods are brown and normally split open (dehisce) longitudinally, releasing the dark brown to black seeds. Good seasons for seed collection are a small mystery with *koaiʻa*. While they flower frequently (most heavily in fall), they do not always set fruit. Determining the reason(s) for this puzzling behavior would certainly be an interesting project.

Germination and Seedling Growth

Treat *koaiʻa* seeds as you would those of *koa*. In one experiment, germination of *koaiʻa* seeds lagged about one week behind that of *koa,* but subsequent seedling growth of the two species was the same. The leaves and stems of *koaiʻa* seedlings are frequently redder than those of *koa* seedlings. After one or two months, when the seedlings reach about 10 cm (4 in) in height, begin to lightly fertilize them to quicken their growth. Like *koa, koaiʻa* seedlings are sometimes attacked and killed by a wilt fungus (probably a *Fusarium* spp.) that initially causes distorted leaves and leaflets. Isolate any seedlings you suspect are infected to prevent the spread of the disease to healthy seedlings. If normal leaf growth does not reappear after repeated spraying with insecticidal soap (for mites that cause similar leaf distortion), destroy the infected seedlings. Pretreating *koaiʻa* seeds with household bleach may help to prevent fungal infection.

Repotting and Planting Out

After they reach a height of 30–50 cm (1–2 ft), plant out *koaiʻa* in full sun following the guidelines in the previous section. Lightly fertilized and well watered, you can expect your *koaiʻa* to grow about 1 m (3 ft) per year for the first couple of years; after that, vertical growth slows. They are highly drought resistant and can go months without watering or rain. (Of course, they do not grow much without water.) Perhaps even more so than with *koa,* there is a great deal of variation in the leaf morphology and color of *koaiʻa*. A grove of *koaiʻa* (grown from seed) is a delightful sight as the grove's genetic variation manifests itself in a multitude of leaf shapes and shades of green. It is less susceptible to leaf damage by Chinese rose beetles than *koa* but, unfortunately, is more prone to infestations of scale insects and mealybugs. Often the scale insects and mealybugs are farmed by a colony of ants at the plant's base. Look for the mealybugs below the soil, attacking the roots. Leafhoppers are also occasionally a pest but are less of a threat than scale insects or mealybugs. Eliminate these pests with repeated sprayings of insecticidal soap or acephate. Eliminate the ants by spreading Diazinon granules around the tree's base.

Kokiʻo
(Kokia drynarioides) — Intermediate

Habitat

Kokiʻo is an extremely rare tree of North Kona's dry forests on the Big Island from about 460 m to 900 m (1,500–3,000 ft). It is federally listed as endangered. We have never seen wild *kokiʻo*—only cultivated plants such as those at Amy B. H. Greenwell Ethnobotanical Garden in Captain Cook, Hawaiʻi.

Hawaiian Uses

Hawaiians produced a pink or lavender dye from *kokiʻo* flower petals (Krauss 1993). They also extracted a dye for fishnets from the bark (Rock 1974).

Whenever we see a *kokiʻo* such as this one at Amy B. H. Greenwell Ethnobotanical Garden, we cannot help but be amazed that so few remain in the wild or that so few people have chosen to cultivate this beautiful cotton tree in their yards. This tree, now 4 m (13 ft) tall and only about nine years old, displays hundreds of blossoms every year, most appearing in summer and fall.

Description

Soon after their notice by Western botanists, these trees, which can reach 10 m (33 ft) in height, were classified in the genus *Gossypium,* which includes commercial cotton and the Hawaiian shrub *maʻo*. But the uniqueness of the Hawaiian cotton trees was soon realized, and they were classified as an endemic genus found nowhere else in the world. While sharing a common ancestor with *Gossypium* species, *Kokia* became known as a spectacular example of evolutionary diversification in Hawaiʻi. In 1853, a New York travel writer on Kauaʻi came upon a wild grove of *K. kauaiensis* at Waiawa, near Mānā, on the *mauka* edge of the island's great southwestern coastal plain. It was June, and the cotton trees in bloom moved the visitor to write, "In this region, vegetation luxuriates in a manner surpassed by few places even in the tropics" (Bates 1854).

Such wonderful endemic displays in Hawaiʻi's lowlands soon disappeared in the wake of sugar plantations and cattle ranches. Just a few patches of cotton trees now remain in the wild, and all face diverse threats: fire, alien insect attack, and aggressive weeds. Of the four recognized species, one, *K. lanceolata,* is extinct. It last grew on Oʻahu's Koko Head in the 1880s. The Molokaʻi species, *K. cookei,* hangs by a thread. It is being propagated by grafting stems of the last individual at Waimea Arboretum onto the rootstock of the two other remaining species. This sole survivor and its grafted (cloned) progeny do not produce viable seed; therefore, horticulturists at Lyon Arboretum are now trying to produce whole plants (stem and root) using high-tech micropropagation techniques. The other two species, *K. kauaiensis* and *K. drynarioides,* are listed as endangered by both the federal government and the State of Hawai'i.

The magnificence of *koki'o* comes in early summer when the trees flower. The blooms are nothing short of spectacular: scarlet, ruffled, and 10–15 cm (4–6 in) wide. A grove of cotton trees in full flower truly can stop travelers in their tracks, as they did in 1853. (Regrettably, today there are no cotton tree groves.) Two thousand blossoms have been reported on a single large tree. The foliage of *koki'o* is also striking—large, thick, glossy leaves shaped somewhat like grape leaves. The wood of sapling-sized plants is extremely flexible; one of the coauthors inadvertently felled a much larger wood tree directly on top of a 3-foot-tall *K. drynarioides,* which bent into a tight U but did not break. The injury in the little *koki'o*'s trunk—crumpling on the concave side and a severe tear on the convex—healed over after one year and did not seem to affect the young tree's growth.

Fruits and Seeds

Koki'o seeds mature in summer and fall. Several large, rounded, wedge-shaped seeds form in the capsule occupying the center of the flower. The seeds are covered with a thick brown fuzz—the cotton. After the petals are lost, the flower's large bracts dry and harden into a radial array of blades. When snapped off in high winds, the whole structure can travel some distance from the parent tree.

Germination and Seedling Growth

Scarify *koki'o* seeds to hasten sprouting. Do so by removing the cotton from at least one area of the seed and then sanding the hard, brown seed coat until the light yellow cotyledons appear through the thinned coat. Avoid sanding all the way through the seed coat; this will damage the seed leaves. Afterward, soak the seeds in water overnight. Germination takes one to two weeks, but older seeds, collected from the ground, seem to take a little longer to sprout than seeds taken from recently matured capsules still attached to the tree.

Sprouts grow quickly; a robust root appears first, closely followed by the shoot, which quickly sheds its seed coat. Large bilobed cotyledons then take one to two days to unfold. Transfer the seedlings to individual pots at this time. The stems of young seedlings soon become uniformly stippled with dark spots of pigment; do not mistake the spots for a disease.

Seedling growth is normally rapid; however, plants in pots may undergo stasis through the summer, perhaps caused by elevated soil temperature (see previous section). Aphids, whiteflies, and red mites sometimes infest the leaves of young *koki'o*. Eliminate these pests by hand or with an insecticidal soap spray.

Repotting and Planting Out

Kokiʻo trees are hardy and do well in a xeriscape. Place your *kokiʻo* in the ground when it reaches 20–30 cm (8–12 in) in height. They thrive in full sunlight but will also tolerate partial shade. When planting out, be sure to provide some shade for the first week or two. Water your *kokiʻo* with big, infrequent "drinks" rather than light, daily showers but make sure the excess water drains away easily. After a year, *kokiʻo* are highly drought resistant, but they grow slowly during the hottest months. Some young specimens we have grown at or near sea level shed many of the leaves during the summer. Our *kokiʻo* have grown about half a meter (20 in) in height per year; a nine-year-old tree at Amy B. H. Greenwell Ethnobotanical Garden on Hawaiʻi is over 4 m (13 ft) tall. *Kokiʻo* have soft bark that is easily damaged by weed trimmers; use a small wire fence or circle of aluminum sheathing to protect young trees.

The thick bark of *kokiʻo* protects it against most chewing pests. Ants may find your plants, however, and begin to farm aphids, mealybugs, or leafhoppers on the younger foliage. This can severely deform the leaves. To intercept the ants, circle the lower stem with a bead of sticky resin (see previous section). With the ants eliminated, the aphids and other insects are attacked by a variety of insect predators and rapidly disappear. If large infestations persist you can treat them with insecticidal soap. Chinese rose beetles sometimes feed on *kokiʻo* leaves, but not as intensely as they do on other large-leafed species such as *maʻo hau hele* or *wiliwili*.

Kokiʻo keʻokeʻo
(Hibiscus arnottianus) — Easy

Habitat

Hibiscus arnottianus live in mesic to wet forest on Molokaʻi and Oʻahu. The Molokaʻi variety, var. *immaculatus,* is federally listed as endangered. A closely related species, *H. waimeae* (also called *kokiʻo keʻokeʻo*), grows on Kauaʻi.

Hawaiian Uses

Hawaiians ate *kokiʻo keʻokeʻo* flower buds as a laxative and made cord from the fiber (Bornhorst 1996). The plant is frequently mentioned in old Hawaiian songs and legends (Neal 1965).

Description

Kokiʻo keʻokeʻo is probably the most frequently propagated native Hawaiian plant, perhaps because of its fragrant flowers (most hibiscus are without fragrance). Many *koki'o ke'oke'o* subspecies and varieties are propagated by stem-tip cuttings in order to perpetuate their distinctive floral characteristics. For example, subspecies *immaculatus* from Moloka'i, now listed as endangered, has a white staminal column and thus is almost completely white. The authors' favorite is a variety called *kanani kea*. Originally from the Ko'olau Range of O'ahu, it is now cultivated by the staff at Lyon Arboretum. Its flowers have large, wide, overlapping petals (see photo).

Propagating *koki'o ke'oke'o* from seed is easy (see below) but involves some risks. The species hybridizes readily with other hibiscus species (reportedly, *koki'o ke'oke'o* were used to develop the first commercially available white hibiscus hybrids). To make sure you have pure seed, pollinate the flowers by hand. One technique is to use an artist's paintbrush to dust pollen on the stigma, the five-part tip of the flower's central column. Then cover the flowers so treated with a bag until the fruits develop and mature. Even then, the natural genetic variation of the seeds is likely to produce plants with slightly different flower or leaf characteristics than their parents. Seeing what new varieties are produced can be fun.

Left unattended, *koki'o ke'oke'o* will grow quite tall—up to 10 m (33 ft)—and tends toward an ungainly shape. We encourage growers to prune their plants to keep them shorter, more attractive (e.g., bushier), and less likely to blow over in a windstorm.

Not only is the native white hibiscus beautiful but—unlike introduced alien relatives—its flowers have a fragrance. Supposedly, the white hibiscus from Kaua'i are the most fragrant. *Koki'o ke'oke'o* come in several varieties; this one, *kanani kea,* is the authors' favorite.

Fruits and Seeds

Ripe *koki'o ke'oke'o* fruits (called capsules) are about 2 cm (¾ in) long, brown, and dry. They often split open to reveal about a dozen fuzzy angular seeds. In cultivation, *koki'o ke'oke'o* flowers and fruits recurrently throughout the year.

Germination and Seedling Growth

Soak *koki'o ke'oke'o* seeds in hot water (about 70°C; 160°F), allowing the water to cool to room temperature; then soak the seeds an additional twenty-four hours in the cool water. Plant the soaked seeds in a tray of clean vermiculite with or without green moss. The seeds will begin sprouting in one to two weeks and continue to sprout for another few weeks. Seeds not pretreated in hot water but soaked for twenty-four hours in cool tap water will also germinate, but sprouting takes longer and is less synchronous. Transplant the seedlings to individual containers after they have one or two true leaves.

Koki'o ke'oke'o seedlings are hardy and grow quickly when lightly fertilized. Expect your young plants to be 10–20 cm (4–8 in) tall in three or four months. Look for aphids or scale insects that sometimes attack the young plants and remove them by hand or treat with insecticidal soap.

You can also propagate *koki'o ke'oke'o* from stem-tip cuttings following the procedure in the previous section. However, because this is a mesic to wet forest plant, we suggest you mist the cuttings more frequently than our prescription of "once a day."

Repotting and Planting Out

Koki'o ke'oke'o requires more soil moisture than *koki'o 'ula* (*Hibiscus kokio* and *H. clayi*) or *ma'o hau hele* (*H. brackenridgei*); therefore, do not let your *koki'o ke'oke'o* outgrow its pot. Repot whenever you find your plant wilting after only a day or two without watering.

Koki'o ke'oke'o can be planted out at any size over 10 cm (4 in). Choose a partially shady or lightly shady location where the soil stays moist. Avoid overfertilizing because this will promote stem growth over root growth, and your handsome *koki'o ke'oke'o* may topple over during a heavy rain or windy day. In about a year, under good conditions, your *koki'o ke'oke'o* will be 1–2 m (3–6½ ft) tall and you will have to start thinking about pruning. By this time, your plant should also be flowering regularly. *Koki'o ke'oke'o*, like *koki'o 'ula*, survive periodic pruning well, but use a sharp shears as hibiscus bark tears easily, promoting disease in exposed stem tissue. Also, we suggest you avoid the unattractive "haircut" pruning so often used on hybrid hibiscus hedges in commercial settings.

Koki'o ke'oke'o have a number of insect pests, including whiteflies, aphids, scale insects, and Chinese rose beetles. Healthy plants normally survive the periodic attacks by these pests. Natural pest predators such as ladybird beetles and spiders often limit or eliminate attacks by the first three pests. Try to encourage the presence of these natural predators in your garden by limiting insecticide use. Planting your *koki'o ke'oke'o* under a window or yard light can discourage Chinese rose beetles (see previous section for other treatments). On two occasions, healthy *koki'o ke'oke'o* have quickly wilted and died in our gardens. We think that a pathogenic soil fungus or nematodes may have been responsible because, when we planted a new, healthy *koki'o ke'oke'o* into the same site, it too quickly wilted and died in less than a week. After the second death, we gave up and planted our third *koki'o ke'oke'o* a few meters away, where it did fine.

Koki'o 'ula
(*Hibiscus clayi* and *H. kokio*)   Easy

Habitat

Hibiscus kokio is uncommon to rare in the dry to wet forests on Kaua'i, O'ahu, Moloka'i, Maui, and perhaps Hawai'i. *H. clayi*, a federally listed endangered species, occurs only in the dry forests of Kaua'i (Wagner et al. 1990).

Hawaiian Uses

The Hawaiians made dyes from *koki'o 'ula* flowers (Krauss 1993).

Description

Koki'o 'ula is a shrub to small tree that can grow to several meters in height. The flowers of *Hibiscus kokio* are quite variable. For example, the subspecies *saintjohnianus*, from Kaua'i, often has attractive orange flowers, while some *saintjohnianus* have dark red flowers and others have bright yellow flowers. (The bright yellow-flowered *saintjohnianus* is a recent discovery that the Lyon Arboretum staff hope to sell soon at their semiannual plant sales.) *H. clayi* flowers are always red.

The leaves of *H. kokio* (and *H. clayi*) also vary in size and shape. A visit to the several gardens and arboreta that host these two species can be confusing because of the sometimes strikingly different cultivated forms. But with all this variety, we are certain you will discover at least one variety of *koki'o 'ula* that strikes your fancy.

Fruits and Seeds

Mature fruits are pale brown capsules. Each capsule can contain one to over a dozen seeds, depending on the pollination success. The seeds are dark brown and about

Is this Hawai'i's state flower—the endangered *ma'o hau hele* or yellow hibiscus? No—this is an even rarer yellow *koki'o 'ula* (*Hibiscus kokio* subsp. *saintjohnsianus*) from Kaua'i. *Hibiscus clayi* is another endangered *koki'o 'ula* from Kaua'i. Its flowers are always red.

4 mm (⅙ in) long. We have never seen *Hibiscus kokio* or *H. clayi* in the wild; cultivated plants flower either continuously or in frequent cycles throughout the year.

Germination and Seedling Growth

Until recently, our *koki'o 'ula* did not produce seed, perhaps because they are not self-fertile. Only after planting two *Hibiscus clayi* next to each other did we begin to observe fruits and seeds. To date, we do not know if self-infertility is a characteristic of the species or only the plants we have grown.

Planted in the standard manner, *H. clayi* seeds begin germinating in three to four weeks. The seedling quickly sheds its seed coat to reveal two large oval cotyledons. Wait until the seedlings have one or two true leaves (in one to two months) before separating and transplanting them to individual containers.

In contrast to our limited experience with seed propagation, we have often propagated *H. kokio* and *H. clayi* from stem-tip cuttings. To do so, follow the guidelines in the previous section. We do not use rooting hormones, nor do we remove or trim any of the upper leaves to reduce transpirational water loss. In fact, in a study conducted by coauthor Koebele and his students, cuttings of *H. clayi* that had a full set of leaves rooted faster than those whose leaves were reduced by half.

Place the potted cuttings in a small covered aquarium or other clear container (we use clear 32 oz plastic cups with lids) under artificial or indirect light and mist lightly once a day. We do not recommend placing the cuttings under a frequent misting system. Speaking with others that have done so, we found they have not been nearly as successful. Within three weeks, nearly all the cuttings should show signs of rooting and leaf growth. After visible rooting (look for roots growing out of the holes in the bottom of the pot) and leaf growth, transplant the new plants into a porous potting medium with a dose of slow-release fertilizer. They should grow quickly.

Repotting and Planting Out

Plant out *koki'o 'ula* (from seeds or cuttings) at any size greater than about 30 cm (1 ft) into full or partial sun. Outplanting will sometimes shock a young *koki'o 'ula* and it will lose most or all of its mature leaves. Do not panic if this happens and inspect the plant daily for new growth. In most cases, the plant will make a successful comeback without special care or treatment.

In lowland areas, an established *koki'o 'ula* does best with occasional watering (once every week or two), unlike its relative *Hibiscus brackenridgei*, which can go for months without watering. Healthy *koki'o 'ula* grow rapidly. One two-year-old plant in one of our more frequently watered gardens has been repeatedly pruned to keep the plant less than 2 m (6-½ ft) in height.

In pots and after outplanting, *koki'o 'ula* are sometimes infested by aphids or whiteflies. Eliminate these pests with three to four sprayings of insecticidal soap. Another outplanting problem is the species' tendency to grow faster above ground than below, causing the plant to topple over in a good wind or after heavy rains. Avoid this by pruning to keep your *koki'o 'ula* short (less than 2 m) or by supporting it with stakes or ties. If it does fall over, do not despair. Promptly right the toppled plant and support it with stakes or ties; we have had to do this on several occasions. Also, less fertilizer will tend to bring root and shoot growth into balance.

Kōlea
(Myrsine lessertiana)   Intermediate

Habitat

There are approximately twenty endemic species of *Myrsine* throughout the main Hawaiian Islands (excluding Kaho'olawe and Ni'ihau). Kaua'i and O'ahu are home to the majority of these. Most *kōlea* live in mesic or wet forests, but a few inhabit dry forests. *M. lessertiana* occurs on all of the main islands except Ni'ihau and Kaho'olawe. It primarily grows in mesic to wet forests but occasionally is seen in fairly dry habitats (Wagner et al. 1990).

Hawaiian Uses

Hawaiians used *kōlea* wood for the posts and beams of *hale* (traditional houses). It was also shaped into anvils *(kua kuku)* on which to beat *wauke (Broussonetia papyrifera)* bark for the making of *kapa* cloth. The Hawaiians made red and black dyes

As pretty as any flower is the bright red new growth of a *kōlea*. These green fruits, attached directly behind the newest leaves, are not yet ripe. Wait until they turn dark and soft before picking them.

from *kōlea,* the former from the sap or bark, the latter from *kōlea* charcoal (Degener 1973). *Kōlea* was also an alternative wood for canoe gunwales, the favored wood being *'ahakea (Bobea* sp.*)* (Krauss 1993).

Description

Depending on the species and probably the habitat, *Myrsine* are either shrubs or small to medium-sized trees. *M. lessertiana* ranges from 2 to 8 m (6½–26 ft) tall and is the most variable *Myrsine* species in growth form and reproductive biology. Most *kōlea* have beautiful, glossy, dark green foliage. The new leaves are often tinted pink or red as they emerge from the branch tips. Away from the immediate seashore, *kōlea* would make an eye-catching specimen plant in most garden settings.

Fruits and Seeds

Ripe *kōlea* fruits are purple or black berries (green when immature) 5–10 mm (3/16–3/8 in) in diameter. They develop in rows on the branches several centimeters behind the terminal foliage, usually from fall through spring. Even outside the flowering or fruiting season you can identify *kōlea* by the rough, tubercular scars on the surface of last year's twig growth. These scars mark the old attachments of the past crop of fruits.

Each fruit contains one seed surrounded by a woody endocarp (the innermost layer of the fruit) and a fleshy, thin layer of pulp. Scrape the pulp away gently—use your fingers. If the fruit is dry, soften it in water for half a day before removing the pulp. The endocarp is thin and brittle and cracking it may threaten the embryo with pathogenic microbes, so we have always sown the intact endocarp with its enclosed seed.

Germination and Seedling Growth

Germination times vary; perhaps there is a correlation with habitat and species. Using standard techniques, *Myrsine lessertiana* from the mesic to wet forest near Volcano, Hawai'i, sprouted in less than one month, while a related species, *M. lanaiensis,* from dry forest in Ka'ū, Hawai'i, took slightly over two months. The root breaks through the thin endocarp seemingly at random and grows slowly, just a few millimeters per week, as the purple or brownish hypocotyl (developing shoot) arches up above the seed. This process can take a month or so until finally the seedling stands up erect 2–3 cm (1 in) high and often with its young leaves still encased in the endocarp. Such seedlings look like tiny mushrooms.

Resist the urge to surgically remove the endocarp from the tiny seedlings. They will shrug it off by themselves, although it may take a month. If you intervene, you will more than likely tear the plant's "head" off and kill it. Also, note that such helmeted seedlings are attractive to curious birds who pull them up, so beware if your plants are outdoors at this time. Several other Hawaiian plants, including *kopiko* and *lumu,* exhibit this same morphological habit. It would be interesting to discover if there is some advantage for these plants that keep their cotyledons covered for a time after sprouting.

Commonly, *kōlea* produces a first cluster of three small leaves, giving it an unusual appearance. The seedling then settles down to generate leaves one at a time in an alternating pattern. Leave the seedlings in the vermiculite seedbed for three or four weeks after these three leaves appear (because the roots grow very slowly) before transferring them to individual pots.

Repotting and Planting Out

Follow standard procedures when repotting *kōlea.* Young, potted *kōlea* are more sensitive to the drying of their soil than many other Hawaiian plants, so do not ignore them. They respond well to moderate amounts of fertilizer, which you can begin giving them after they are 10 cm (4 in) tall. The new foliage, as the plant grows larger, becomes ever more attractive with its rosy tint.

So far we have not successfully transplanted *kōlea* into the ground. Coauthor Culliney attempted to establish several young *Myrsine lessertiana* in a suburban garden near Volcano, Hawai'i. These plants grew slowly for several months but then ceased growing and slowly declined in health, eventually dying after about one year. Possible causes for the failure were drought (when they were not watered) and competition with mature *'ōhi'a lehua* trees.

Ko'oko'olau
(Bidens amplectens) Easy

Habitat

Bidens amplectens is restricted to the windward cliffs of the northern Wai'anae Range on O'ahu, but there are eighteen other endemic species of *ko'oko'olau* in Hawai'i. Together their combined habitats span the range from coastal bluffs through dry, mesic,

Hawaiians made a hot tea from the leaves of *koʻokoʻolau*. This species, one of about nineteen endemic to our Islands, is now found naturally only in the Waiʻanae Range of Oʻahu but grows equally well in Hawaiʻi's lowlands.

and wet forests to subalpine woodland. Individual species distributions are often but not always restricted to one habitat (Wagner et al. 1990).

Hawaiian Uses

Hawaiians used the leaves of several *koʻokoʻolau* species in hot teas and tonics to treat a variety of ailments. Today many people still use the leaves to make an herbal tea (Krauss 1993). The alien weed *(Bidens pilosa)* is also sometimes incorrectly sold as a "Hawaiian tea."

Description

Bidens amplectens is a small (usually less than a meter tall) perennial, or sometimes annual, herb that is somewhat woody at its base. The plant has pinnately compound leaves and large (4–8.5 cm across, about 2 in), bright yellow flowers. Actually, each "flower" is a composite flower head of many smaller individual flowers surrounded by one set of floral rays. *B. amplectens* has perhaps the largest flower head of all the native *koʻokoʻolau* (Wagner et al. 1990).

Botanists think that all nineteen species of endemic *koʻokoʻolau* evolved from a single common ancestor. While the species differ in morphology and habitat, studies by Ganders and Nagata (1984) have shown that they are all interfertile (i.e., they can hybridize). In some locations you can find natural hybrids, such as those between *B. amplectens* and *B. torta*. Often, however, the species are distinct because of non-overlapping distributions.

The ability of *koʻokoʻolau* to hybridize should concern Hawaiian gardeners, particularly if they intend to sell or give away seed from their plants. Keep records of where your plants or seeds came from. Also, prevent hybridization in your garden by resisting the temptation to grow more than one species at a time. If you want to grow a new species, pull up and discard the old plant or wait a couple of years and let it die naturally. Finally, when selling or giving away seed, insist that the recipient practice the same responsible behavior.

While these precautions may seem like a lot to ask, the alternative is "native" gardens filled with unnatural hybrids, where none of the *koʻokoʻolau* are truly Hawaiian. Hawaiʻi's botanists also worry that these hybrids might mix with wild populations, altering the unique genetic identities of the various species. You can help by never planting out your cultivated *koʻokoʻolau* in wild spaces—which, by the way, is also against the law.

Fruits and Seeds

The fruit of *koʻokoʻolau* is referred to as an achene. Many develop on each flower head, turning gray to black (sometimes dark brown) when ripe. *Bidens amplectens* has achenes about 1 cm (⅜ in) long and straight. They lack the hooks found on the achenes of some other species such as the introduced weed called beggartick *(B. pilosa)*. These hooks help the achene attach temporarily to passing animals, thus dispersing the alien species widely. *B. amplectens* flowers in the late fall and winter; look for ripe achenes in spring and summer.

Germination and Seedling Growth

Plant the achenes horizontally and slightly below the surface of fresh vermiculite in a seed tray. Do not cover the vermiculite with moss because many of the germinating seeds lack the strength to force their way through. Achenes will begin germinating in one to two weeks, pushing their long, thin cotyledons above the vermiculite. Normally not all germinate, so it is best to plant more than you want. Seedling growth is quite rapid, with four to six leaves appearing in two to three weeks. Transplant the seedlings at this stage to individual containers.

Repotting and Planting Out

Bidens amplectens should be repotted as it grows. Left in a small pot, the plant becomes stunted and may not flower. When planting out, choose a location with full or partial sun. Water frequently (two or three times a week) until the plant begins to grow vigorously, after which you can reduce the watering to one to four times a month, depending on the location. As the plant grows, old leaves on the stem normally turn brown and die but do not fall off right away. You may wish to remove these by hand to keep your *koʻokoʻolau* attractive. To date, we have limited experience with this species when planted out and therefore cannot comment extensively on its longevity or pest problems. So far the *koʻokoʻolau* we have planted have grown very quickly, with stems growing as fast as 10 cm (4 in) a week. They began to flower in the fall, about two months after planting out. We have not seen any signs of disease or pest damage.

Kōpiko (*Psychotria hathewayi* and *P. hawaiiensis*) — Intermediate

Habitat

Many of the eleven *kōpiko* species in Hawaiʻi are found with *kōlea, lama, olopua,* and *ʻōhiʻa lehua* in mesic forests from foothill elevations to about 1,500 m (5,000 ft). Some *kōpiko,* such as *Psychotria hathewayi,* endemic to the Waiʻanae Range of Oʻahu, and *P. hawaiiensis,* found on Molokaʻi, Maui, and Hawaiʻi, also inhabit dry forests (Wagner et al. 1990). When choosing a *kōpiko* for your garden, pick a species that naturally grows under conditions similar to those in your garden.

Hawaiian Uses

Hawaiians fashioned *kōpiko* wood into *kapa* anvils (Lamb 1981).

Description

Kōpiko are a group of eleven endemic *Psychotria* species in the coffee family (Rubiaceae). All are attractive shrubs to small or medium-sized trees. *Psychotria hathewayi* is a small tree up to 8 m (26 ft) tall, although the plants we have grown, now over three years old, are still less than a meter high. *P. hawaiiensis* can grow to be a tree 12 m (40 ft) tall but, again, it takes many years to achieve this height (Wagner et al. 1990).

Some *kōpiko* can be distinguished by their *piko*—small depressions on the undersides of the leaves, usually near the vein axils. (The Hawaiian word *"piko"* means navel or other small swelling.) They are also recognizable by the tightly pressed together pair of young leaves that emerge at the tips of branches. *Kōpiko* produce short stalks of tightly clustered, small, white or yellow flowers.

Alien birds such as bulbuls have discovered that the ripe, red fruits of this *kōpiko* are a tasty treat. The two plants here began flowering and fruiting after less than a year in the ground and have continued to do so ever since.

Fruits and Seeds

The availability of *kōpiko* seeds is highly variable depending on species and climate. The fruits are pea sized and turn yellow, orange, or red (or purple in some species) when ripe. Typically two brown to whitish yellow, hemispherically shaped seeds occupy most of the fruit's volume. These seeds have their flat inner surfaces pressed together (like those of *lama*) so that they form a composite pit. You can easily remove the ripe pulp with your fingers and clean the seeds by rubbing them under a stream of water.

Germination and Seedling Growth

Without scarification, *kōpiko* seeds sprout a white root within two or three weeks of sowing. Although the roots emerge rather quickly compared to many Hawaiian forest trees, shoot growth is quite slow. The seedlings commonly remain 1–2 cm (less than an inch) high with their cotyledons enclosed in the seed's husk for one or two months or longer. Finally, growth resumes and the cotyledons expand as they split the walls of the seedcoat. The *kōpiko* seedling now resembles a tiny nasturtium because of the semicircular seed leaves.

In *Psychotria hathewayi* and *P. hawaiiensis,* seedling growth is slow to moderate; after six months, seedlings are about 10 cm (4 in) tall. On occasion, we have had seedlings very quickly wilt and die (within two or three days) even though the soil was moist. We suspect this may be the result of overwatering. Aphids and scale insects sometimes infest young *kōpiko*. Remove the pests by hand or kill them with an insecticidal soap spray.

Repotting and Planting Out

We have had no problems repotting *kōpiko*. Our experience in planting it out is limited to two *Psychotria hathewayi*. We planted these out in a garden that is irrigated three times a week and receives full sun for about half the day. The plants have done well, growing in three years to nearly a meter in height. They began flowering and fruiting after less than a year in the ground and have continued to do so since then; at least one of the two plants always has flowers or fruits or both. Beneath the plants, several seedlings have sprung up through the lightly mulched soil. The abundant African snails *(Achatina fulica)* in the garden do not seem interested in the *kōpiko*, nor do the Chinese rose beetles. We have never seen either species on the two plants. On two or three occasions, however, we have removed a grasshopper from the shrubs and suspect that the limited leaf damage is likely due to these insects.

Lama
(Diospyros sandwicensis) Intermediate

Habitat

Wild *lama* occur in dry to mesic forests and occasionally in wet forest from near sea level to about 1,200 meters (4,000 ft) on all the main islands except Kahoʻolawe and Niʻihau (Wagner et al. 1990). These trees grow in a wide range of soil types and thrive exposed to full sunlight as well as moderate shade. On the Big Island, *lama* commonly grows on older lava with plants such as *ʻākia (Wikstroemia sandwicensis), alaheʻe (Psydrax odorata), ʻohe makai (Reynoldsia sandwicensis),* and *wiliwili (Erythrina sandwicensis)*.

Hawaiian Uses

Inside a *hālau hula,* a building for dancers and trainees, a block of *lama* wood, wrapped in yellow, scented *kapa,* was placed on a *kuahu* (altar). It represented both Laka, the goddess of hula, and enlightenment—the word *"lama"* means light (Abbott 1992). The Hawaiians also used *lama* wood to construct sacred *hale* (houses) within a *heiau* (temple). Other traditional uses of *lama* included eating the ripe fruits after first drying them, making a poultice for treating skin sores from pulverized *lama* wood and other ingredients, and building fishtraps with *lama* branches (Krauss 1993).

Description

Lama is still a fairly common tree in less disturbed habitats, even on Oʻahu. It is an attractive tree because of its striking black bark, colorful fruits, and dense foliage whose new growth is often pink or rose colored. The wood of *lama* is dark, heavy,

The colorful ripe fruits of *lama* taste somewhat like a persimmon. This young *lama* tree shows off its namesake with some bright pink new leaves. *Lama* means "light" in Hawaiian.

and fine grained; it has been called "Hawaiian ebony." Woodworkers report that *lama* sawdust looks and feels like volcanic sand (Keith Zeilinger, pers. comm.). A distinctive field character of *lama* is its leaves; they commonly grow in nearly the same plane, resembling multiple pairs of wings sprouting from the branch.

There are two species of *lama*: *Diospyros hillebrandii* and *D. sandwicensis*. *D. hillebrandii* can be distinguished from *D. sandwicensis* by its larger, dark green leaves that, on their upper surfaces, have a reticulate-pitted pattern; *D. sandwicensis* leaves are normally lighter green and lack this textured upper surface. *D. hillebrandii* occurs only on Kauaʻi and Oʻahu; it is common along the lower stretches of the Awaʻawapuhi Trail in Kokeʻe State Park, Kauaʻi. *D. sandwicensis* grows on all the main islands except Kahoʻolawe and Niʻihau. We have grown only the latter species.

Fruits and Seeds

Lama fruits resemble small persimmons (the species are closely related) and turn yellow, orange, or red as they ripen. The fruits are edible but rarely sweet. Persimmon fanciers will recognize the astringent taste of *lama* fruits. Inside the fruit is a large, composite, brownish pit that typically splits longitudinally into two or three sections. Each wedge-shaped section is a seed. A single tree normally bears fruit only once a year, but individual trees—depending on the island, weather, and elevation—will bear fruit at different times throughout much of the year.

Germination and Seedling Growth

Lama seeds sprout readily two to four weeks after sowing. To monitor the sprouting, keep the seeds on the surface of the vermiculite seedbed under a moist layer of green moss. The root emerges from the more pointed end of the seed.

Initial growth is dominated by the development of a long, dark, wire-stiff taproot for the first one to three weeks. This is followed by shoot growth featuring broad, green, oval cotyledons that may persist for three weeks or more before the appearance of the first true leaves. The cotyledons themselves remain enclosed in the seed coat for some time—a sprouting behavior shared with other Hawaiian plants such as *kōpiko* and *kōlea*. It is best to leave such "helmeted" seedlings alone; attempts to remove the seed coat often result in decapitating the seedling. Eventually the seedling will shrug off the seed remnants.

Pot *lama* sprouts after the first phase of root growth, whether or not the cotyledons have emerged. Use a tall pot to accommodate the seedling's long taproot but be sure that water can penetrate all the way through the soil in the pot.

In the early stages of shoot formation, *lama* seedlings face two severe threats. The first is a tiny unidentified borer that attacks and destroys the growing tip of seedlings kept outdoors. Even seedlings enclosed by window screen died as victims of this pest. This borer is less common in winter and spring and may be a localized threat. (The problem has been severe in Waimānalo but absent in Waipahu/Pearl City.) The second threat is a stem-rotting fungus on young *lama* seedlings that resembles that seen on similar-aged seedlings of *kauila*—both *Alphitonia* and *Colubrina*. (Refer to the entry on *Colubrina* for an effective treatment.)

Repotting and Planting Out

Lama do not like pots. If not planted in a deep pot, they develop a severe coiling of the taproot, a condition also seen in other Hawaiian plants such as *'a'ali'i* and *uhiuhi*. This is a problem when the plants are planted out. In the worst cases, the plant literally strangles itself in two or three years as the roots thicken. If you see this problem and you want to save the plant, try to induce new root growth just above the coil. Do this by scraping the outer bark gently in three or four places and applying rooting hormone (obtainable at garden supply stores). Repot the plant and wait several months, then check for new root growth. If there are new roots, cut off the old root mass at the top of the coil. When planting out such a specimen, prune the foliage more or less severely until the root growth catches up to meet the water demands of the plant.

Even in deep pots *lama* grow slowly or not at all. This seems to be true even if the plants are adequately watered and given fertilizer. We therefore recommend planting your *lama* out when it is still quite small—less than 20 cm (8 in). Initially the young seedling does not grow much; perhaps it is in shock or most of the growth is in the roots. After a couple of months in the ground, however, the small tree starts putting out new leaves and branches at a moderate pace. After two or three years, healthy *lama* saplings can grow as much as half a meter (20 in) in height per year.

In cultivation as in nature, *lama* thrive in xeric to mesic settings with well-drained soil, in full or partial sun, from sea level to over 1,000 m (3,300 ft). They grow in both volcanic and calcareous soils. One five-year-old *lama*, planted in a backyard near Waimānalo Beach, Oʻahu, is 3 m (10 ft) tall.

A tiny unidentified insect will attack and kill the growing tips of new *lama* branches. These attacks occur primarily in the spring (see previous section under stemborers). A few other pests such as scale insects, a small gray weevil *(Myllocerus sp.)*, and the Chinese rose beetle attack *lama* from time to time, but the tree's growth is only marginally affected by these pests (see previous section for treatment of these pests).

Loulu
(Pritchardia remota) — Easy

Habitat

Pritchardia remota occurs naturally only in the valleys of Nihoa, a small ancient island northwest of Kauaʻi (Wagner et al. 1990). It is federally listed as endangered. Cultivated trees are present on all the main islands.

Hawaiian Uses

The Hawaiians fashioned spears from the hard wood of some *loulu* and used the leaves for thatch. They also ate the seeds. To appease fishing gods, the Hawaiians built seasonal *heiau* made of *loulu* fronds (Abbott 1992). Today the leaves are woven into hats and fans (Lamb 1981).

Description

Pritchardia remota is one of nineteen species of fan palms endemic to the Hawaiian Islands. It is closely related to three other species: *P. aylmer-robinsonii* on Niʻihau, *P. glabrata* on Maui, and *P. napaliensis* on Kauaʻi (Wagner et al. 1990). These four species, like the other fifteen, are each endemic to one island within the Hawaiian chain (i.e., no *loulu* species occurs naturally on more than one island). Some have suggested that Hawaiian gardeners should restrict themselves to growing only those *loulu* species endemic to their island. While there is some sense to this (as a precaution against hybridization), we disagree. Often your front or backyard will not match a local species' preferred habitat, and the chance of hybridization between a residentially grown *loulu* and the wild trees is virtually nil. Additionally, today some *loulu* species are quite rare and occur in only one small area. A fire or other disturbance could potentially exterminate the species. We recommend growing those *loulu* species with a preferred habitat similar to your residence (e.g., dryland *loulu* for leeward lowland yards, wet forest species for upper-elevation gardens). We also encourage you to keep good records of the species you are growing and not plant different

A beautiful young *loulu (Pritchardia remota)* at Queen Kapiʻolani Garden near the Honolulu Zoo. Protected from rats, this tree still has many ripe fruits.

loulu species too close to one another. If planted close together, the chance of producing hybrid seeds is great—something we should all actively avoid.

Fruits and Seeds

Pritchardia remota fruits are nearly spherical and about 2 cm (¾ in) in diameter. When ripe, they are dark brown or black. The fruit has three layers: a shiny outer skin called the exocarp, a fibrous middle layer called the mesocarp, and a hard inner endocarp. There is one oily seed inside each fruit. It is attached to the endocarp at one small spot called the hilum. *Loulu* seeds remain viable for a long time; fruits collected over two years ago still contained seeds that germinated and grew well.

Germination and Seedling Growth

There are two methods of seed preparation for *loulu;* both are successful but the second results in more rapid germination. The first method is to remove both the exocarp and mesocarp of the fruit and soak the hard inner endocarp and contained seed in water for one or more days prior to planting. The second method is to remove the exocarp, mesocarp, and most of the endocarp and then plant the seed immediately. To remove the hard endocarp, carefully crack it with a pair of pliers or vise (actually, we use a small C-clamp for better control) and remove it in pieces by hand. Be careful not to damage the seed or remove the attached portion of the endocarp at the hilum.

Do not plant the seeds in the standard vermiculite bed—it is too shallow. Instead, plant each seed in a deep but small pot containing either a 1:1 mix of vermiculite and cinder or a 1:1:1 mix of vermiculite, perlite, and peat. Germination takes four to five weeks for the second method and somewhat longer for the first. After the first one or two leaves appear, fertilize the seedling to hasten its growth. Plants four to six months old are about 30 cm (1 ft) tall and have three to five leaves. These leaves are entire and only after one to two years do they begin to split and assume a fanlike appearance.

Repotting and Planting Out

Loulu should be repotted frequently or planted out to ensure continued growth. *Pritchardia remota* responds well to frequent fertilization and watering, putting out a new leaf about once a month. In particularly dry areas, our outplanted *loulu* have grown poorly or died. After two years, expect the palm to be about 1 m (3 ft) tall. After three to four years, the *loulu* will be 2–3 m (6–10 ft) tall and a supportive skirt of roots begins to grow out from the lower trunk.

Except for the occasional grasshopper and a scattering of whiteflies, few insects have attacked the leaves of our *loulu;* however, an unidentified stem-borer did kill one of our young plants. This palm began putting out small deformed and damaged leaves and later, leaf production stopped completely. Unfortunately, none of the insecticides (insecticidal soap, acephate, pyrethrin) used to combat the insect worked. We finally removed and destroyed the palm after several months of no growth.

Another potential problem is chewing damage from unidentified arthropods at the palm's base. This caused one of our palms, about 2 m (6½ ft) in height, to fall over during a strong windstorm. Surprisingly, this tree survived; it rerooted itself after being propped up and secured in place. Granular Diazinon, sprinkled around the base several times per year, can effectively control these pests.

Maiapilo
(Capparis sandwichiana)   Difficult

Habitat

Maiapilo is a coastal plant, but it occasionally grows inland in dry areas. Known as "Hawaiian caper," it occurs on all the main islands as well as Laysan Island, Midway Atoll, and Pearl and Hermes Atoll (Wagner et al. 1990).

Hawaiian Uses

While capers are the pickled flower buds of *Capparis spinosa,* a relative of *maiapilo,* we could find no recorded uses for this plant.

Description

Maiapilo is a low-lying shrub with light green leaves, large white flowers, and small cucumber-like fruits. The flowers open in the early evening, last through the night,

Maiapilo is also known as the "Hawaiian caper." The large white flowers open in the early evening, releasing a strong, pleasant fragrance. By midmorning the next day, they turn pink and begin to wilt.

and then wilt by midmorning. They have a strong, pleasant fragrance. Flowering is most prevalent during the spring and summer months. It is likely that native moths originally pollinated *maiapilo* because of its night-flowering habit.

Fruits and Seeds

Maiapilo fruits resemble small cucumbers, initially green and then turning orange yellow when ripe. The ripe pulp is bright orange and soft. Fruits are ripe and seeds mature in late summer through fall. The size of the fruit is quite variable. We have seen small fruits 3–5 cm (about 1½ in) long on *maiapilo* in North Kona, Hawai'i, and considerably larger fruits 6–8 cm (about 2½ in) long on *maiapilo* from Barbers Point, O'ahu. Urban birds such as bulbuls *(Pycnontus cafer* and *P. jocosus),* after learning the ripe orange pulp and seeds are edible, frequently "steal" the fruits or hollow them out, leaving the husks on the stem. Full-size but still green fruit will ripen at home by keeping the stem of the fruit in water. Change the water and recut the stem every one or two days to prevent mold and keep water flowing to the fruit. This technique will not work if the fruit is still growing in size, so resist the temptation to pick young fruits. The fruit contains many reddish brown to black seeds, each approximately 3 mm (⅛ in) in size, that resemble a tiny coiled snail shell.

Germination and Seedling Growth

Using our standard procedure, *maiapilo* seeds sprout quickly (one to three weeks) and usually with high success (50–95 percent). On two or three occasions, however, all the seeds from a fruit failed to germinate; we do not know why. Transfer the seedlings to individual containers after two or three true leaves develop (one to two months). Most seedlings grow rapidly after transplanting, while some wilt and die quickly (from a damping-off fungus?) and still others languish and die for no apparent reason. Young plants are victims to a host of chewing pests, such as caterpillars, cockroaches, various beetles, slugs, snails, mice, rats, and even rabbits. These pests often eat the entire seedling.

Repotting and Planting Out

Maiapilo varies in its response to repotting. While most of the time it does well, it will sometimes lose all its leaves. After this it usually puts out new growth, but sometimes the plant just dies. The varied response may be related to root damage during transplanting, so be careful. Planting out success is also variable; some plants die within one or two weeks while others begin growing, often quite rapidly. In any case, plant *maiapilo* out in full sunlight. After three or four months, the plant should require little watering. *Maiapilo* begin flowering at a small size and young age (less than six months).

Unfortunately, even large plants are still attacked by chewing pests, most notably caterpillars, slugs and snails, and rodents. To curb insect attack, try using a systemic insecticide such as acephate (e.g., Isotox at 2 tbs per gallon). Usually attacks on large plants are not fatal but limited to one or two branches. We have also seen damage by an unidentified stem-borer. This borer eats out the stem's interior, causing the distal leaves to quickly wilt. Finally, even healthy *maiapilo* have the unnerving habit of unexplained stem death. The leaves on the stem will turn yellow, wilt, and die over a period of days or weeks. The stem tissue then dies, turning yellow and brittle. Following the death of the stem or concurrent with its death, the plant will put out new growth either at the base of the stem or elsewhere on the plant.

Despite all the above problems, it is our experience that mature *maiapilo* are quite resilient. Given the beauty of their flowers, they are a worthwhile challenge for the Hawaiian gardener.

Maile
(Alyxia oliviformis) — Easy

Habitat

Maile occurs most frequently in mesic forests but also in wet and dry forests and shrubland, from elevations as low as 50 m (170 ft) to about 2,000 m (6,600 ft), on all the main islands except Kahoʻolawe and Niʻihau (Wagner et al. 1990). The driest place we have seen *maile* is in Kaʻū, Hawaiʻi, where it grows on old *ʻaʻā* lava as part of a sparse understory along with *ʻūlei, ʻalaʻala wai nui,* and bracken ferns; overhead grow *lama, ʻōhiʻa lehua, ʻiliahi,* and *Alphitonia* trees.

Hawaiian Uses

Hawaiians have long fashioned the fragrant *maile* into leis, traditionally as garlands (open chains) of leaves and other decorations. *Maile* was one of the five essential decorations placed on the *kuahu* (altar) within a *hālau hula,* a building for dancers and trainees, consecrated to the goddess Laka (Handy and Handy 1972). There it represented the four Maile sisters, mythical sponsors of hula. Hawaiians also used *maile* to scent *kapa* (Abbott 1992).

Quite out of its natural habitat, this potted *maile* appears content in the partial shade of a *hala* tree in Wai'anae. *Maile* fruits look like olives when ripe—hence the name *oliviformis*.

Description

Hawaiians recognized the highly variable leaf form of *maile* and assigned names to at least five of these forms: *maile lau li'i*—small-leafed maile; *maile lau nui*—big-leafed maile; *maile pākaha*—blunt-leafed maile; *maile ha'i wale*—brittle maile; *maile kaluhea*—sweet maile (Wagner et al. 1990). Today, the large-leafed varieties, sometimes grown outside Hawai'i, are frequently favored by commerce (Bornhorst 1996). When the leaves are crushed or the outer layer of the vine stripped or peeled away, *maile* releases a sweet fragrance resembling vanilla. The foliage itself is deep green and glossy. Together, fragrance and appearance account for the attractiveness of leis made from this plant.

Today, at numerous public ceremonies such as school graduations, *maile* leis are traditional gifts, and as a result our native forests suffer. Even when the *maile* is harvested in the traditional manner, too many collectors threaten the declining wild populations. Careless collectors damage other forest plants as well, as they trample them underfoot while gathering *maile* off established trails. The obvious solution (we believe) is for lei makers to begin growing their own *maile* in backyards or community lei gardens. It is encouraging to note that a few community lei-making groups have already begun this practice.

While *maile* appears vinelike, botanists formally classify it as a shrub. With its extensive branching and elongation, it can spread many meters from its base and climb high into large trees. Tiny white and yellow tubular flowers, 2–3 mm (⅛ in) in diameter, appear in small clusters. They have a faint vanilla-like fragrance.

Fruits and Seeds

Maile fruiting occurs mainly in fall and winter. The fruits are usually singular and resemble olives or slightly elongate grapes. Occasionally you will see a multiple-seeded fruit; they look like two or three single *maile* fruits fused together in a line. The fruit is initially green, turning deep purple or nearly black when ripe. The seed is a large football-shaped pit. When you remove the fruit's pulp surrounding the seed, you will encounter a milky, slightly sticky sap. The sap is easily washed off the seeds with water, but it may not come off your clothes as easily.

Germination and Seedling Growth

Using our standard method, *maile* seeds sprout readily in two to four weeks without scarification. A prominent white root appears first, followed by a long thin shoot that reaches 3–5 cm (1–2 in) high before a tiny pair of leaves unfolds at the tip. Initial growth of *maile* can be slow, but it accelerates after one to two months. Newly potted plants tend to grow as single shoots and may reach a meter in height after six months.

Repotting and Planting Out

You can grow *maile* in a pot to quite a large size (2 m in height) provided you give it a large pot (1–2 gallons), but we encourage you to plant it in the ground to optimize its growth. They tolerate planting out well but may be sensitive to soil type or companion plants. For example, some *maile* grown on the Big Island, propagated from seeds collected in old-growth mesic forest, grew poorly in younger soil near the parent plant's habitat even though the younger landscape supported mature *'ōhi'a lehua, koa,* and other native plants.

Most *maile* prefer partial shade, although the Ka'ū variety grew best in full sun (with watering three times a week). Flowers on our plants appeared in their second year of growth. Coauthor Culliney is growing *maile* at sea level in a slightly raised garden where the soil is partially calcareous, while coauthor Koebele has plants thriving in heavy clay soil. The plants at both sites have grown at about 30–50 cm (12–20 in) per year.

Pests that attack *maile* include scale insects and aphids, normally attended by ants. Eliminate the scale insects and aphids either by hand on a small plant or with two or three sprays of insecticidal soap. Bad infestations of scale insects may require a combination of manual and chemical intervention. Systemic insecticides such as acephate are also useful in combating large populations of scale insects.

Māmane
(*Sophora chrysophylla*)   Easy

Habitat

Māmane grows on all the main Hawaiian Islands except Ni'ihau and Kaho'olawe. It occurs in dry to mesic forests on most islands and at subalpine elevations on Maui

Winter and spring are the time to watch your *māmane* turn yellow with hundreds of pealike flowers. Afterward, the fruits ripen and usually stay attached for many months, making seed collection an easy task.

and Hawai'i (Wagner et al. 1990). The cultivation notes below refer to a shrubby dryland variety growing at about 500 m (1,600 ft) on Hawai'i.

Hawaiian Uses

Hawaiians used *māmane* wood for house construction, for *'ō'ō* (digging sticks) and sled runners (Lamb 1981). They also used the wood for spears (Abbott 1992) and strung *māmane* flowers into leis (Krauss 1993). In nineteenth-century Hawai'i, *māmane* wood was used for fence posts and the spokes, axles, and frames of wagons (Lydgate 1883).

Description

Māmane is a legume tree in the same family as *koa* (Fabaceae), although it does not grow as fast or tall. The wood is tough, flexible, and rot resistant. These trees produce feathery compound leaves and, in winter and spring, bloom with clusters of bright yellow flowers. In regions where native honeycreepers such as *'apapane* and *'amakihi* live, naturalists have observed that flowering *māmane* trees attract these birds (Lawrence Katahira, pers. comm.). *Māmane* trees on the Big Island's Mauna Kea are also the primary food source for the endangered *palila*, another honeycreeper with a finchlike beak that is used to crack the tough *māmane* seeds (Hawai'i Audubon Society 1981).

Fruits and Seeds

The *māmane* pods, characteristically winged and constricted between the seeds, stay attached to the tree for most of the year. They are green when immature and brown or brownish gray when ripe. Often viable seeds are abundant on the ground under the trees; the yellow or orange seed coat is highly rot resistant.

Germination and Seedling Growth

Scarification (vigorous scratching or sanding a small area of the seed coat), followed by a day's soaking in water, is essential for speedy germination of *māmane* seeds, which takes one to two weeks. (The scarified seeds color the water a bright yellow orange.) *Māmane* seeds sprout in an unusual manner that initially resembles *maile* germination. A tall thin shoot emerges from the seed shortly after the root appears. The tiny first set of leaves is inconspicuous; no exposed fleshy cotyledons emerge in this bean family plant. These are soon followed, within one to two weeks, by the first compound leaves. Young *māmane* grow vigorously but should be protected from intense direct sunlight, strong winds, and heavy rain until they are about 20 cm (8 in) tall.

Repotting and Planting Out

Māmane are hardy and are easily repotted or planted out into the landscape, where they need lots of sunshine to thrive. One fast-growing seedling planted into a suburban yard near Volcano on the Big Island grew to 2.5 m (8 ft) high and flowered in four years; however, others grew much more slowly. Like *koa*, *māmane* are hosts to nitrogen-fixing bacteria (see the Nitrogen Fixing Tree Association website: www.winrock.org/forestry/factnet.htm), and the variable growth rates of the plants may reflect the state of the symbiosis between a particular *māmane* and its complement of microbes in the soil. If you are planning to grow *māmane* in an area where they do not naturally occur, you might try to inoculate your seedlings with a small amount of soil from beneath a healthy wild or cultivated tree.

In the wild, *māmane* rarely occurs at altitudes less than 450 m (1,500 ft), but we have successfully planted out the Big Island's dryland variety near sea level (30 m; 100 ft). After three years, these *māmane* are nearly 2 m (6½ ft) tall but a bit spindly. Plants in pots at sea level generally grow poorly, perhaps because the soil temperature is too high.

Māmane is susceptible to a small thrips that chew away the bases of the small leaflets, giving the tree a somewhat bare appearance. Chinese rose beetles also initially attack the leaves of *māmane,* but after the first defoliation, subsequent attacks are much less common. It would be interesting to investigate how the plant achieves this later resistance.

Ma'o
(Gossypium tomentosum)   Easy

Habitat

Ma'o grows in dry shrubland from the coastal zone to low-elevation foothills (to 120 m; 400 ft). It is curiously absent from Hawai'i but occurs on all the other main islands

Hawaiians made dyes for *kapa* (bark cloth) from both the leaves and flowers of *ma'o*. Today, this Hawaiian cotton is more commonly used as an ornamental in coastal gardens.

(Wagner et al. 1990). *Ma'o* is reportedly extinct in the wild on Kaua'i (National Tropical Botanical Garden 1992). Fire or urban and agricultural development has destroyed most of the *ma'o* habitat so that today wild plants are much rarer than in the past. Still, it persists in some fire-disturbed sites such as Makapu'u and Wai'anae, O'ahu.

Hawaiian Uses

Hawaiians made dyes for *kapa* cloth from *ma'o*. Modern *kapa* maker, Dalani Kauihou, makes a bright, long-lasting green dye by immersing *ma'o* flowers in hot water (pers. comm.).

Description

Some *ma'o* grow as upright shrubs over a meter tall. Others are low-lying shrubs less than half a meter tall (20 in), with branches extending 2 m or more along the ground. One variety from Ni'ihau, now probably extinct, was a small tree (pers. comm. from K. Robinson to K. Woolliams). Heredity undoubtedly is responsible for some of this variation in form, which is readily apparent when you see *ma'o* native to different regions grown together. But the environment can also determine its shape, such as in windy sites where *ma'o* grow close to the ground. *Ma'o* leaves resemble those of maple trees but have a feltlike surface. The normally abundant flowers, which appear for several months from late winter to summer, are sulfur yellow and about 5 cm (2 in) in diameter. The silvery leaves and bright flowers together make *ma'o* a most attractive plant for lowland Hawaiian gardens.

A member of the large plant family Malvaceae, which includes *'ilima* and hibiscus, *ma'o* is sometimes called "Hawaiian cotton." It is a close relative of the commercial cotton plant, also classified in the genus *Gossypium*. While the cotton fibers of *ma'o* are relatively short and unsuitable for processing into thread, the plant is resistant to many of the diseases and pests that plague commercial cotton. The cotton industry has exploited this resistance by cross-breeding *ma'o* with commercial cotton to produce disease-resistant commercial hybrids (Arrigoni 1978). This is a good example of why we need to save our native plants. While many native plants may not be commercially valuable themselves, their relatedness to important commercial crops could make them priceless in the future.

Fruits and Seeds

Ma'o fruits are small cotton bolls, brown and dry when mature from late spring through fall. The fruit contains as many as six wedge-shaped seeds, each about 6 mm (¼ in) long and densely covered with brown cotton fibers.

Germination and Seedling Growth

Using our standard method, *ma'o* seeds begin sprouting quickly, often in as little as two or three days. The seedling has large bilobed cotyledons; these are soon followed by the first true leaves, which are triangular. Later leaves take on the maple leaf appearance. Sucking insects such as aphids and whiteflies sometimes infest *ma'o* seedlings. Eliminate these either by hand or with an insecticidal soap spray.

Repotting and Planting Out

We have transplanted *ma'o* seedlings directly from the seedbed into the ground when the plants had just a few true leaves and were only 10 cm (4 in) tall. These young plants need to be partially shaded and carefully watered for a few weeks, but they will then thrive in full sunlight. You can also, of course, transfer the seedlings to individual pots and plant them out at a later time and larger size.

Ma'o does well in a wide range of volcanic to calcareous soils, provided there is good drainage. Even poor-looking stony ground, with modest fertilizing, will produce a verdant growth of *ma'o*. It is an excellent plant for xeriscapes. After about a year, when the shrubs are a meter or more in diameter, they will require little or no watering. Expect to see flowers and fruits within the second year of growth.

Ma'o is a coastal or dryland plant and does poorly in upland areas or overwatered gardens. Under these suboptimal conditions, a leaf-spotting fungus *(Colletotrichum sp.)* will often infect *ma'o* (National Tropical Botanical Garden 1992). If your shrub becomes infected, remove the spotted leaves and reduce the amount of watering to control the disease. You can also try a fungicide, although we have never resorted to this treatment. Upland planted or overwatered *ma'o* sometimes lose their leaves and die quickly. Pathogenic nematodes or fungi in the soil are likely causes of these deaths.

The most serious pest we have encountered growing *ma'o* is the cotton aphid, *Aphis gossypii*. This insect attacks the new growth at the tips of branches. As the young leaves grow, the aphids cause them to become stunted and twisted, forming ugly knotlike tangles. Like many other aphids, this species is farmed and protected by ants. Rarely life threatening to a healthy plant, the infestation results in unsightly patches of foliage for up to several weeks at a time. Eventually the aphids seem to disappear, only to return after a few months. You can treat infestations by carefully removing the worst of the twisted leaves and spraying the affected branch tips with insecticidal soap or acephate. If you can identify the routes by which ants gain access to your plants, you can block their paths with a sticky resin (see previous section). Controlling the ants seems to clear up the aphid problem quickly.

Other insect pests on *ma'o* include the treehopper *(Vanduzeea segmentata)*, Chinese rose beetle, whiteflies, and aphid species other than the cotton aphid. Most often we have ignored these pests because either the damage did not warrant the time and energy for treatment or natural predators eliminated the pests for us. For more immediate treatments, however, refer to the previous section.

Maua
(Xylosma hawaiiense)   Intermediate

Habitat

Maua grows primarily in mesic forest but also in dry and wet forest on all the main islands except Ni'ihau and Kaho'olawe (Wagner et al. 1990). Cultivation notes below refer to a dryland variety from North Kona, Hawai'i.

Hawaiian Uses

The Hawaiians likely used the dense wood of *maua* to make poi-pounding boards (Lamb 1981).

Description

Maua is a smallish tree (3–9 m; 10–30 ft tall) with attractive shiny leaves. New leaves and stems are burgundy red (similar to many *lama*). The long petiole (leaf stem) makes the leaves flutter in the wind. Early in this century, field botanist Joseph Rock lamented the few remaining *maua* trees he found in the arid coastal hills of west Moloka'i as "the remnant of what was once a beautiful forest" (Rock 1974). Today, *maua* is rarer still—on all the islands.

Maua is usually dioecious, meaning that some trees have flowers that produce only pollen while others have flowers that, when pollinated, develop fruits. Many other Hawaiian plants are dioecious, such as the common *'a'ali'i (Dodonaea viscosa)* and the very rare *mēhamehame (Flueggea neowawraea)*. Knowing if a plant is dio-

The abundant and colorful fruits of this *maua* tell you it is a female tree because this species is usually dioecious (individual trees have either male or female flowers).

ecious or monoecious (a plant with flowers that produce both pollen and fruits) can be important when planning a garden. If your intention is to produce fruits, perhaps to give away to friends, you need only one plant if it is monoecious—but you will need at least two if it is dioecious.

Fruits and Seeds

Maua berries are deep red when ripe and about 1 cm (⅜ in) in diameter. Each fruit contains a few small, irregularly shaped, black seeds. Look for the ripe berries in summer and fall.

Germination and Seedling Growth

Using our standard procedure, germination is good and takes three to five weeks. Seedlings grow rapidly after transferring them to individual containers and, in as little as two months, they can be 20 cm (8 in) or more in height. One problem with such rapid growth is that the seedling falls over and must be supported. Avoid this problem by moving the young seedlings early into an open location where the trunk and stems will thicken and stiffen in response to the wind.

Repotting and Planting Out

Maua responds well to repotting and if not continually repotted will grow slowly. Plant it out in an area with full or partial sun when it is 20–30 cm (8–12 in) tall.

After you see new growth, begin to lightly fertilize the tree to maintain its rapid growth rate (about 1 m per year). *Maua* grows well in a variety of soil types, including soil with a high clay or calcareous sand content. Less drought tolerant than most of the other dryland trees included in this chapter, *maua* may require supplemental watering for several years in dry areas.

Chinese rose beetles sometimes chew on *maua* leaves, but not often. New leaves and stems occasionally become "burned," wilting and turning black. It is unclear if this is a response to drying, excessive sun or fertilizer, or some unobserved pest or disease. The alien black twig-borer *(Xylosandrus compactus)* can be a serious pest to young *maua*. We have seen the most damage under mesic to wet conditions, less in dry areas. Early and thorough spraying of the borer's holes (in the stem) with an insecticide such as malathion or acephate will kill this pest and sometimes save the stem. If the stem is already dead, cut it off immediately and destroy the borers by hand. Do not leave infested stems near any of your plants.

Naio
(Myoporum sandwicense) — Easy

Habitat

Naio has an extremely wide distribution, from coastal areas through dry, mesic, and wet forests. It also occurs in subalpine forest. It grows on all the main Hawaiian Islands except Kahoʻolawe (although it probably once inhabited this island as well) (Wagner et al. 1990).

Hawaiian Uses

Hawaiians used the hard wood of *naio* for net gauges and house posts and frames (Krauss 1993). Near the end of the sandalwood boom in Hawaiʻi (the trade ended after 1845), *naio* was sometimes substituted for true sandalwood *('iliahi)* in shipments to China and elsewhere. Yet while the spicy odor of fresh *naio* heartwood is similar to that of *'iliahi*, shipments arriving in China were normally rejected by buyers. From this commerce, *naio* acquired the unflattering name "bastard sandalwood" (Arrigoni 1978).

Description

Naio is usually a shrub or small tree, but occasional mountain specimens can be as much as 12 m (40 ft) tall. Botanists have divided the species into three groups: subspecies *st. johnii* found in Kona, Hawaiʻi; variety *stellatum* found in ʻEwa and Barbers Point, Oʻahu; and subspecies *sandwicense,* found throughout the main islands (Wag-

Naio varies greatly in stature as well as leaf, flower, fruit, and branching characteristics. Here branches from two *naio* (placed side by side) show how the flower color and petal number can vary between individual plants living in the same area. On the Big Island, a *naio* crawls along the rocky coastline, quite unlike its more common upright relatives inland.

ner et al. 1990). While dividing the species in the above manner helps to categorize some of the morphological variation seen in this species, considerable variation within each group is also present. For example, in the variety *stellatum* there are shrubs with small white flowers and shrubs with large purple-throated flowers; shrubs that have pink-spotted flowers and shrubs with nonspotted flowers; shrubs with long thin leaves and others with short wide leaves; and shrubs with frequent branching and shrubs with long uninterrupted branches. Some of this variation is probably environmentally induced, but part of it is also genetic, such as the spots on the flowers (based on experiments by coauthor Koebele). With so much variability within one population, we wonder if natural selection has gone on vacation (in Hawai'i) when it comes to *naio*.

Fruits and Seeds

The ripe fruits of *naio* are fleshy, slightly less than 1 cm (⅜ in) in diameter, and normally white in color, although occasionally a shrub will have light pink or purple fruits; green fruits are unripe. Some *naio* produce fruits continuously throughout the year; others have definite periods when flowers and fruits are plentiful and periods when they are not. Even these latter shrubs normally fruit more than once a year. Within the fleshy fruit is a hard capsule (the endocarp of fruit) that contains several chambers or cells where the small white or whitish yellow seeds reside.

Germination and Seedling Growth

After cleaning the capsules of their fleshy fruit, randomly sample a few to verify that the *naio* you have collected from is consistently producing capsules with at least one fully developed seed. Do this by scraping or sanding the apex of the capsule (i.e., the end farthest from the original stem attachment) to expose the chambers. Use a magnifying glass to examine the chambers. Often, most if not all of the chambers will contain undeveloped seed, in which case the chamber will appear empty (the undeveloped seed is small and flat). In contrast, a fully developed seed normally fills its chamber and, if separated from the capsule, looks like a tiny white sausage.

Unless you are quite careful when scraping to expose the chambers, you will damage the good seed within these capsules. If you do, the seed will normally not germinate and you should discard it. After verifying that the *naio* is producing good seed, gently scrape or sand the apex of each remaining capsule (not nearly as severely as you did to verify good seed) and place the capsules in clean tap water for one to three days to soak. Replace the water once or twice daily. Plant the capsules in a tray of clean, moist vermiculite, burying them about 1 cm (⅜ in) below the surface. Do not place the capsules directly on the surface of the vermiculite. This will cause problems later during germination. In contrast to the standard procedure, do not cover the vermiculite with green moss. The moss in this case seems to increase the likelihood of bacterial or fungal infection of the seeds, perhaps because the seeds remain too wet. Keep the vermiculite in the tray moist by gently watering or misting when necessary.

After about two weeks, some seeds will begin to sprout. The root of the seed emerges first from a hole in the apex of the capsule. Later, the stem and cotyledons pull themselves free of the capsule and rise above the surface of the vermiculite. Sometimes a seed will germinate but fail to free its leaves from the capsule, particularly if you plant the capsules too shallow. Leave these seedlings to free themselves rather than attempt the delicate operation of separating the capsule from the new leaves; this nearly always fails. If you have good seed, most will germinate by the end of two months.

The above technique has worked very well for fruits from the *stellatum* variety of *naio* but not nearly as well for the more common subspecies *sandwicense*. Scraping the capsule's apex is not necessary for germination, but experiments by coauthor Koebele did show that it decreases germination time and increases germination success. After seven weeks, over 80 percent of the scarified capsules had germinated, compared to 30 percent for nonscarified capsules.

Keep the new *naio* seedlings under bright light to prevent spindly growth. After the first set of true leaves appears in two to four weeks, transfer the seedlings to individual containers and begin light fertilization. Seedlings grow quickly and you should have a sizable plant—20 cm (8 in) tall—in about six months.

Growing *naio* from seed is an extremely rewarding endeavor. While it is somewhat of a challenge, the directions above should help you succeed. But if you have tried and failed or if you do not feel you are yet up to the challenge, there is an easy alternative method of propagating *naio:* from stem-tip cuttings. To grow *naio* from stem-tip cuttings, follow the general directions given in the previous chapter. This method is particularly useful if you want to produce a number of shrubs all with the same morphology, perhaps for a garden hedge.

Repotting and Planting Out

Naio tolerate repotting quite well, but it is best to eventually plant them out. Plant out anytime after they have reached 20–30 cm (8–12 in) in height for best success. As stated earlier, *naio* have an extremely wide climatic range, so we encourage you to choose a location similar in temperature, moisture, and light levels to that of the parent plant's location. Most of the plants we have grown came from parents living in a hot, dry, and sunny habitat. When planting out in a location similar to this, it is best to initially shade the plant and water it whenever there are signs of drought stress (two or three times a week for the first month or so). Our plants have grown quickly to over a meter (3 ft) in height and width in one to two years (the parent plants were not much larger than this). After six months to a year, *naio* can normally survive on their own without supplemental watering.

Naio is without question one of the most pest resistant native plants available to the Hawaiian gardener. Perhaps because of its sap, few of the typical pests (e.g., Chinese rose beetles) will chew on it, although we have seen grasshoppers feed on *naio* leaves in one extremely dry coastal setting. On rare occasions, scale insects or aphids show up on some of the stems or leaves, but they have always disappeared without treatment. Also, on occasion a stem tip will die, we think because of an unobserved stem-borer, but this has never become serious and has never killed a plant. Some growers have told us that *naio* is sometimes attacked and killed by a root pest—perhaps a nematode or fungus—but we have never had this happen.

Naio is also one of the few native plants that can hold its own in competition with alien grasses. The soil beneath the larger specimens we have planted out is free of grasses, even though these aliens surround the shrubs.

In light of its pest resistance, ability to compete successfully with alien grasses, variety of forms to choose from, and ease of propagation, we hope that more and more people in Hawai'i will decide to plant this attractive shrub around their homes.

Nama sandwicensis
(no Hawaiian name)   Easy

Habitat

Nama grows in sandy soil or raised limestone reef rock as part of the strand or coastal vegetation. It occurs on Lisianski, Laysan, and all the main islands except Kaho'olawe (Wagner et al. 1990).

This mature *Nama* is less than 20 cm (8 in) across and 3 cm (about 1 in) tall. Every October, a new batch of *Nama* seedlings appears in this planter filled with coral sand and rubble. In just a few months, they will mature, flower, and set seed.

Hawaiian Uses

Hawaiians do not seem to have used this plant.

Description

Nama sandwicensis is the perfect plant for the lover of small things. It is an extremely small, matlike plant less than 5 cm (2 in) tall, growing to a maximum diameter of about 30 cm (1 ft). It has tiny, fuzzy, and somewhat succulent leaves that curl under at the margins. The flowers too are small (about the same size as *naio* flowers) and vary in color from blue to white with a yellow throat. *Nama* is reported to be either an annual or a short-lived perennial, but our plants have never lived even a year before senescence (Wagner et al. 1990). Endemic to Hawai'i, *Nama*'s closest relative lives in the southwestern United States (Carlquist 1980).

Fruits and Seeds

Nama has a capsule fruit 3–4 mm (⅛ in) long that contains many diminutive seeds. Collecting the seeds directly from a plant is nearly impossible because of their small size. Instead we recommend collecting the leaf and flower litter, along with a very thin layer of sand, from beneath a mature plant. This mix will contain many seeds.

Germination and Seedling Growth

Sprinkle the mix of litter and sand onto the surface of a pot filled with washed beach sand and coral rubble. Water gently to bury the tiny seeds in the sand, then place the pot outside in full sun or in a very sunny window and water about once a week. *Nama* germination seems to depend on the right combination of day length and soil moisture. Unfortunately, we cannot be more specific about what constitutes a "right combination" except to report that repeatedly in the month of October, a large number of *Nama* seeds germinated in an outdoor planter that was watered regularly. This suggests that a shortening day length might be the trigger for *Nama* germination. In any case, your patience will be rewarded.

Seedlings are tiny but also grow very fast. Do not fertilize the seedlings or mature plants as the fertilizer will either kill the plant or cause it to grow vinelike, rather than in its natural matlike form.

You can also propagate *Nama* from stem-tip cuttings using the standard technique described in the previous section, but we discourage this propagation method because it is not as successful as growing from seed. Under no circumstances, despite the temptation, should you dig up a wild plant and take it home. This is not only illegal, but *Nama* is far too uncommon for this type of behavior.

Repotting and Planting Out

Nama normally do not need to be repotted. If you wish to grow them in a garden rather than a pot, you should follow the above procedures. Be certain the garden area receives full sun and that the soil is very sandy. Once established in a pot or a garden, *Nama* will reseed itself. Typically, there will be a month or two when all the plants die off; then one day you will discover a new set of seedlings. If after six months you do not get a new set, you will need to collect fresh seed from the field (or perhaps from a friend who is also growing *Nama*). Insect pests on *Nama* appear limited to aphids, mealybugs, and scale insects. These are not common but if present can be easily eliminated with two or three treatments of insecticidal soap. Be certain to eliminate the ants that usually bring in these pests.

Nanea
(*Vigna marina*)   Easy

Habitat

This Hawaiian beach pea grows mainly over strandline dunes and coastal flats and sometimes colonizes rocky hills and cliffs near the sea. On the main Hawaiian Islands, only Lānaʻi, Niʻihau, and Kahoʻolawe do not appear to have wild *nanea* populations (Wagner at al. 1990).

Hawaiian Uses

The Hawaiians do not seem to have used this plant.

Nanea makes an attractive groundcover, or it can be trained to grow over garden walls and fences.

Description

A legume vine, the *nanea* has attractive bright green foliage and produces many yellow flowers from winter to summer. This plant typically hugs the ground but given the opportunity will climb over low shrubs and fences. It is fast growing and may tend to take over your garden; from time to time *nanea* usually needs selective pruning.

Fruits and Seeds

Seeds are khaki-colored beans, nearly spherical, and lentil sized. Up to several occur in each pod. The seeds have a thick coat and require scarification for timely sprouting.

Germination and Seedling Growth

Seedlings are hardy, grow rapidly, and transplant easily, but they are attractive to some pests, including the devastating bean fly *(Ophiomyia phaseoli)*, which deposits eggs on *nanea* stems. The tiny fly larvae burrow into the tissue and hollow out the stem, killing everything above the damaged region. In seedlings, the favorite place for this fly to lay its eggs is near the soil line, almost always killing young plants. Where the bean fly is abundant, as it is in much of Windward Oʻahu, you should grow young *nanea* indoors or outdoors under fine screen.

Repotting and Planting Out

Nanea is easily transplanted, and once the plants are a meter long and branching, with older stems more than 4 mm (3/16 in) thick, they usually survive attack by the bean fly. Another insect that may attack older *nanea* is the small black stinkbug, *Coptosoma xanthogramma*. Do not mistake this pest for the beneficial black lady beetle that it resembles. The stinkbug sucks sap from the stems of plants it attacks. *Nanea* usually survive attack, although stems are scarred by long infestations. Slugs and snails sometimes find the foliage of *nanea* palatable.

Nanea makes an attractive interplanting and ground cover among native trees and upright shrubs such as *wiliwili, hala pepe, lama,* and coastal *ʻākia,* whose foliage is generally sparse enough to let much of the sunlight through. It will tend to overgrow low-growing shrubs such as *ʻōhai* and coastal *ʻiliahi. Nanea* usually flowers in its second year.

Nānū
(Gardenia brighamii)   Easy

Habitat

Nānū was probably common in Hawaiʻi's ancient dry forests. Now it is extremely rare in the wild, with perhaps a couple dozen trees scattered over five islands—Hawaiʻi, Lānaʻi, Maui, Molokaʻi, and Oʻahu. Fortunately, many of Hawaiʻi's botanical gardens and others have cultivated this native gardenia. *Gardenia brighamii* is federally listed as an endangered species.

Hawaiian Uses

Hawaiians beat *wauke (Broussonetia papyrifera)* to make *kapa* on an anvil sometimes made of *nānū* wood. The *kapa* was then stained yellow using dye made from ripe *nānū* fruit. Fragrant *nānū* flowers were often strung into leis (Krauss 1993).

Description

Gardenia brighamii is the smallest of three species of native gardenia, all called *nānū* by the Hawaiians. Like Hawaiʻi's more common introduced gardenias, our native *nānū* has shiny dark green leaves and fragrant white flowers. While the other two species, *G. mannii* and *G. remyi,* are trees, *G. brighamii* is usually a shrub or in some places, such as Kānepuʻu, Lānaʻi, a small tree. Since most of us in Hawaiʻi have small yards, if we have any yard at all, this is good news.

Undoubtedly one reason for its rarity in residential areas is that *G. brighamii* is

Until recently, it was against the law to have this beautiful native gardenia, called *nānū* by the Hawaiians, in your yard without a permit. Thankfully, that law has been changed. The fruit below the flower in this photograph is still immature; wait until it is soft before picking it.

listed as an endangered species by both the federal government and the State of Hawai'i. Until very recently, this meant that it was illegal for most people to propagate or even have this attractive plant around their homes. Fortunately, our state law was changed in 1998 (owing to the hard and persistent work of many concerned people), and now you can legally buy and possess our native *nānū*. Licensed growers, such as Lyon Arboretum and Foster Gardens, are permitted to sell "surplus" plants to the public. You as a buyer in turn must agree not to plant your new possession near any wild *nānū*. Doing so might spread a fatal disease or cause changes in the genetic makeup of the wild population. But then, why would you ever want to plant your *nānū* in the forest when it looks so much nicer in your front yard?

Fruits and Seeds

When ripe, *nānū* fruits are about the size of a golf ball, green with small white spots, and have a soft yellow pulp. Each fruit contains many yellow, disk-shaped seeds about 5 mm (³⁄₁₆ in) in diameter. *Nānū* fruits take their time ripening on the tree. You can test to see if a fruit is ripe by squeezing it; a ripe fruit is soft and your fingers will stain as you penetrate the bright yellow pulp. If the fruit is still hard, leave it on the tree because the unripe fruit and its seeds will not mature properly when prematurely picked.

Germination and Seedling Growth

Remove the seeds from the ripe fruit and clean them thoroughly of any attached pulp. Place them in a glass of tap water for about an hour; viable seeds will sink to the bottom and you will see the water turn yellow as the seeds release their pigments. Do not sterilize the seeds with bleach; rinse them and plant them directly into a bed of clean vermiculite with or without green moss. If you plant them without moss, bury them about 5 mm (3/16 in) below the surface. If you are not quite ready to plant your *nānū* seeds, it is better to leave them in a ripe fruit rather than clean them and store them dry. We have kept ripe fruits for two to three weeks and still had excellent germination of the seeds. Take care, however, to not let the ripe fruit become moldy.

The seeds will begin sprouting in about two weeks and continue for another two or three weeks. Some seedlings take one or two weeks to shed their seed coats; avoid the temptation to "help" them. Seedlings have a long white taproot, so it is better and easier to transfer them from the vermiculite to individual containers about a week after they have shed their seed coats. After the seedlings have one or two pairs of true leaves, you can begin watering them with a dilute fertilizer such as Miracle Gro. Young seedlings are prone to attacks by mites that cause distorted leaf growth, and if ignored, they can kill the seedling. If you see evidence of mites, or even as a preventative treatment, spray the seedlings three or four times (three to four days apart) with a half-strength solution of insecticidal soap. In around three months your seedlings should be about 10 cm (4 in) tall and you can begin thinking about where you are going to plant them out.

Repotting and Planting Out

Young *nānū* are robust plants that tolerate repotting well, but wait until they are 20–30 cm (8–12 in) tall before planting them out. Smaller plants, although capable of surviving the transition from pot to ground, are more likely victims of hungry snails, birds, and mammals. *Nānū* do best in either full sun or partial shade, but for the first couple of weeks we recommend you provide daytime shade so that the plant acclimates to its new conditions. Look for signs of new growth to confirm that your *nānū* has successfully acclimated.

While *nānū* seedlings grow quite quickly, we have seen slower growth with them once they are planted out—perhaps 20–30 cm (8–12 in) per year. Because the places we have planted out are quite hot and dry, however, this growth rate might not be typical for more sheltered or more frequently watered plants. After reaching 1–2 m (3–6½ ft), *nānū* begin to flower and bear fruit. At this size they are highly drought resistant and need watering in only the driest of settings.

Adult *nānū* are resistant to all the common pests (e.g., Chinese rose beetles, aphids, stem-borers), and we have never had to intervene with any type of treatment. Occasionally fire ants set up a nest at the base of a *nānū,* perhaps to feed on the flowers' nectar. Use the commercially available insecticide hydramethyinon (AMDRO) to eliminate them.

Naupaka
(Scaevola coriacea and S. gaudichaudii) Intermediate

Habitat

Scaevola coriacea is federally listed as endangered and is currently limited to a few coastal areas on Maui and Moloka'i. It formerly occurred on all the main islands except Kaho'olawe. *S. gaudichaudii* grows on dry ridges, open shrubland, and forest on all the main islands except Ni'ihau and Kaho'olawe (Wagner et al. 1990).

Hawaiian Uses

The fruits of some *naupaka* were used to make a black dye, and they were also occasionally eaten (Krauss 1993).

Description

Scaevola coriacea is a remarkably prostrate shrub, almost vinelike in appearance, with succulent oval leaves and small white flowers. *S. gaudichaudii* is a low-lying shrub with attractive, deep-throated, yellow orange flowers. Both species are in the same genus as the ubiquitous beach *naupaka (S. sericea)*. The beach *naupaka* is easily grown from either seed or cuttings and is used extensively in coastal landscaping.

The flowers of all Hawaiian *naupaka* (there are nine species recognized by Wagner et al. 1990) are unusual; all the petals extend from one side of the flower's throat.

The prostrate dwarf *naupaka*, Scaevola coriacea, is a good choice for sand-filled planters and coastal or rock gardens. In contrast, choose the more upright dryland *naupaka*, S. gaudichaudii, with its deep-throated, yellow orange flowers, for a border, low hedge, or accent plant.

This arrangement makes the flower appear as if it has been torn in half. There are several Hawaiian legends regarding *naupaka*'s unusual flowers. One legend (Neal 1965) tells of a beautiful woman who fell in love with a young village man, but he did not love her and returned to his village sweetheart. The beautiful woman followed and soon expressed her anger by tearing the man away from his lover's embrace. Only then did the two realize that the beautiful woman was actually Pele, the goddess of fire. Pele pursued the man into the mountains, hurling rocks at him. The other gods, pitying the man, transformed him into the half-flower of the mountain *naupaka*. Pele was furious and, on a river of lava, returned to the coastline where she overtook the man's sweetheart. But before she could harm the young woman, the gods turned the woman into the half-flower of the beach *naupaka*. Today, if you wish to unite the two lovers, you must hike into the mountains and pick a flower from the mountain *naupaka,* bring it down to the seashore where the beach *naupaka* grows, and gently press the two flowers together.

Fruits and Seeds

The fruits and seeds of *Scaevola coriacea* and *S. gaudichaudii* are similar. The fruits are ovoid, 5–10 mm (about ¼ in) long, and dark purple or black when ripe. Each contains one large yellowish white seed. Mature plants of both species bear ripe fruits periodically throughout the year.

Germination and Seedling Growth

We have had very poor success in germinating *Scaevola gaudichaudii* using the standard procedure described in the previous section and only slightly better success with *S. coriacea*. The limited germination has been as rapid as one month and as slow as nine months. In some recent experiments, however, we soaked the seeds (of *S. gaudichaudii*) in a 0.05 percent solution of gibberellic acid for two days prior to planting, which resulted in about 60 percent germination over a period of three to six weeks. (You can buy gibberellic acid from a biological supply company such as Carolina Biological Supply, Inc.)

Propagation from stem-tip cuttings is also possible. Use cuttings of mostly new, green wood and follow the standard procedure. Rooting takes one to three months and is about 50 percent successful. Seedling and rooted cutting growth is good, and you can expect small (10–20 cm; 4–8 in) but healthy plants in about six months. During this stage and later, *S. gaudichaudii* grows more rapidly than *S. coriacea*.

Repotting and Planting Out

Both species tolerate repotting well, but when planting out recall the plant's habitat. *Scaevola gaudichaudii* should be planted in full sun in dry areas; after the plant is established it will require little or no watering. In contrast, we have planted out *S. coriacea* only in large planters filled primarily with sand, some cinder, and natural clay soil. We have, however, seen successful plantings of *S. coriacea* in noncoastal areas (e.g., Liliʻuokalani Gardens, Oʻahu).

S. gaudichaudii grows rapidly after outplanting, becoming more than a meter (3 ft) in diameter after a year or less. *S. coriacea* grows much more slowly. Both species have responded well to dilute foliar fertilizer (e.g., Miracle-Gro). Scale insects occasionally infest *S. gaudichaudii* but are eliminated with insecticidal soap. More commonly, spittle bugs infest the numerous stem tips. Although they seem to cause little or no damage to the plant, the spittle bugs can be eliminated with two or three treatments of insecticidal soap. In contrast to *S. gaudichaudii,* the most common pest of *S. coriacea* is spider mites that attack primarily the young leaves; left unchecked, they can seriously threaten the entire plant. The mites can be controlled or eliminated, again, with two or three treatments of insecticidal soap. While the soap does not kill a mite-infested *S. coriacea,* it will often cause the mite-damaged leaves to turn brown and die.

S. gaudichaudii is one of the few low-growing Hawaiian native plants we have grown that is able to successfully compete with alien grasses such as *Chloris barbata.* Despite neglect, two of our plants surrounded by alien grasses have continued to flourish for over two years.

ʻOhe makai
(Reynoldsia sandwicensis) Intermediate

Habitat

This increasingly rare tree grows in the dry forests of Niʻihau, Oʻahu, Molokaʻi, Lānaʻi, Maui, and Hawaiʻi (Wagner et al. 1990).

Hawaiian Uses

The Hawaiians made stilts called *kukuluaeʻo* (for games and play) from the wood of *ʻohe makai* (Krauss 1993). They also used the resin from cut branches for unspecified applications (Rock 1974).

Description

ʻOhe makai can be a tall tree (possibly 30 m; 100 ft, indicated by Wagner et al. 1990) but most we have seen are considerably smaller (less than 10 m; 33 ft). It has nearly white bark, white and soft wood, and flexible, sometimes twisted branches. During the summer it normally sheds its compound leaves, each of which is composed of

The early morning sun reflects off an *'ohe makai* tree in North Kona on the Big Island. Shortly before bankruptcy, the owner of the land where this tree grows was considering developing the area into a golf course. *'Ohe makai* seedlings start life with simple single leaves. As they grow older they develop compound leaves with three, then five, then seven leaflets.

seven large leaflets. You can easily spot an *'ohe makai* tree from a distance because of the contrast between its light green leaves and the usually darker green of the surrounding flora.

Fruits and Seeds

'Ohe makai normally flowers at the beginning of the fall rains. The fruits are small, 7 mm (¼ in) in diameter, occur in clusters, and turn purple and soft when ripe. Ripe fruits are most plentiful in winter and spring. Within each fruit are several soft, flattened seeds. Fresh, viable *'ohe makai* seeds sink in freshwater. (The rule, "sinkers grow, floaters don't," is true for the seeds of many Hawaiian plants, such as *alahe'e, āulu, hao, kauila,* and others.)

Germination and Seedling Growth

Extract the seeds from the ripe fruit and wash them in tap water to remove any remaining pulp. Plant them directly in a clean bed of vermiculite; do not disinfect the seeds in household bleach. You can also dry the seeds for planting later but if you do, soak them for twenty-four hours before planting. Initially all the dry seeds will float, but after soaking viable seeds will sink. Germination takes two to six weeks. Germination of seeds stored more than three months is noticeably less than that of fresh seeds.

'Ohe makai seedlings initially put out several simple leaves. Later, when the plant is about 15 cm (6 in) tall, the new leaves begin emerging in a three-leaflet form and then later still in the mature seven-leaflet form. Seedlings quickly develop a large fleshy root immediately below the soil surface. In the wild, the seedling undoubtedly stores water in this root in order to survive its first few dry summers.

If watered too frequently, a seedling will lose its leaves; if overwatering continues, the seedling will die. Seedlings sometimes become infested with aphids, scale insects, mites, or whiteflies. Look for abnormal leaf development or premature leaf death as evidence of these pests and eliminate them with two or three sprayings of insecticidal soap several days apart.

Repotting and Planting Out

Our experiences with *'ohe makai* are limited. To date, we have germinated seeds and grown seedlings that after about ten months have reached a height of approximately 30 cm (1 ft). We have also planted out more mature plants (about 1 m in height) given to us. Unfortunately, all these older plants have met with unnatural ends; one was run over by a truck, another girdled by a careless weed whacker. However, while the plants were in the ground (for about two years), in full sun, growth was slow. Over the summers, the plants would lose all their leaves and become dormant. Over the winters, neither increased watering nor fertilizer seemed to hasten the plants' growth. Our observations of plants in other areas, such as Koko Crater, suggest that young *'ohe makai* saplings may initially grow better under more shaded conditions. While growth rates were disappointing, pest resistance was encouraging. Our plants were ignored by chewing insects such as caterpillars or Chinese rose beetles. Occasionally, scale insects (brought in and protected by ants) would infest one or more leaves, but this was easily controlled with either insecticidal soap or acephate.

'Ōhelo kai
(Lycium sandwicense)   Easy

Habitat

In Hawai'i, *'ōhelo kai* grows in rocky coastal areas on all the main islands. It also occurs outside Hawai'i on Rapa, Tonga, and the Juan Fernandez Islands (Wagner et al. 1990).

The extremely drought resistant *'ōhelo kai* is another good candidate for a coastal or rock garden. Unlike the sweet fruits of *'ōhelo (Vaccinium)*, the tomato-like fruits of *'ōhelo kai* would make a poor-tasting jam or pie filling.

Hawaiian Uses

The Hawaiians apparently did not use this plant, although the name reflects the somewhat similar appearance of the fruits to those of *'ōhelo (Vaccinium)*.

Description

'Ōhelo kai is in the nightshade family, as is the tomato. This spreading shrub, normally less than half a meter tall, produces many small white and blue flowers in the fall and winter. Its succulent leaves and bright red berries make it an attractive groundcover for sandy or rocky gardens.

Fruits and Seeds

'Ōhelo kai fruits are bright red when ripe, normally during the fall and winter, and nearly 1 cm (⅜ in) in diameter. Within each fruit are many (a dozen or more) tiny, flattened seeds, resembling those of a tomato. Viable seeds sink in water and can be separated from most of the fruit pulp by crushing the fruit in a container of water; the viable seeds will fall to the bottom of the container.

Germination and Seedling Growth

Prior to planting, soak the seeds in water containing a very small amount of household bleach; use just a few drops of bleach per pint (500 mL) of water. After fifteen minutes, use an eyedropper to transfer the seeds to a freshly prepared, well-moistened seed bed of vermiculite. Do not cover the seed bed with moss as the extremely small seeds and seedlings may become entangled in the moss fibers. Instead, cover the seed bed container with wax paper about 2 cm (¾ in) above the vermiculite, keep it in the shade, and watch for sprouting within a week.

After sprouting, remove the wax paper. The seed bed's moisture will begin to evaporate more quickly, and you must compensate for this. One successful technique with this species and others having very small seeds is to place the seed bed container in a shallow pan of water for one or two minutes each day. Perforations in the bottom of the seed bed container allow water to be wicked up through the vermiculite to the rapidly growing roots of the plants. By watering from the bottom, you also avoid damaging the tiny seedlings and discourage fungal infection of the stem and leaves.

Seedlings grow quickly and should be transplanted to individual containers when they have four to six true leaves. The plants will continue to grow quickly in their own containers with little or no fertilization. After they reach 5–6 cm (2 in) in height, gradually acclimate the young plants to full sunlight. When moving your plants outside, beware of snails and slugs that may make a midnight snack of your *'ōhelo kai* seedlings.

Repotting and Planting Out

'Ōhelo kai tolerate repotting well. After the plant reaches 20–30 cm (8–12 in) in height or length (as the plant soon begins growing horizontally), plant it out in a location with full sun. *'Ōhelo kai* does best in sandy, rocky soil similar to that of its normal habitat. Plants that are overwatered, overfertilized, in the shade, or planted in soil containing an abundance of clay will be more spindly than wild plants, and their leaves will be less succulent. If you want your plants to look like those in the wild, water and fertilize sparingly and keep the plant in full sun and sandy, rocky soil. If your garden soil differs from this, you may want to consider planting your *'ōhelo kai* in a large planter with sand, coral rubble, and cinder.

Within a year of planting out, expect your plant to be about half a meter (1½ ft) in diameter (remember, this is a plant that grows low to the ground with infrequent branching). Also within the year, it should begin to flower and set fruit.

Sucking insects such as aphids, mealybugs, scale insects, and whiteflies sometimes infest *'ōhelo kai*. Unless the infestation becomes extensive, however, natural predators such as ladybugs normally eliminate these pests over time. If not, treat the infested plant with four or five sprayings of insecticidal soap over a two- to three-week period.

Pāpala kēpau
(Pisonia brunoniana, P. sandwicensis, and P. umbellifera)   Easy

Habitat

In Hawai'i, native *Pisonia* are primarily found in inland mesic habitat, especially along the sides of streams, from 100 to 1,500 m (330 to 5,000 ft). The closely related

Hawaiians used the sticky gum exuded from ripe *pāpala kēpau* fruits to catch small forest birds prized for their colorful feathers. This large *Pisonia umbellifera* has leaves nearly 30 cm (1 ft) long.

Pisonia grandis, however, grows along the seashore on other tropical Pacific islands such as Samoa (Whistler 1980).

Hawaiian Uses

Hawaiians used the sticky resin (birdlime) exuded by *pāpala kēpau* fruits to catch small forest birds prized for their colorful feathers. The feathers were woven into fancy cloaks, headdresses, and leis. After a bird catcher found a population of targeted birds, he would strip the nearby shrubs or small trees of their leaves, then smear the gluelike substance over the branches. The birds, apparently curious about the leafless shrub, would land on the sticky branches and become glued to the spot, whereupon they were easily captured (Degener 1973).

Description

Depending on the species, *pāpala kēpau* range from large shrubs to small trees up to 15 m (50 ft) tall; *Pisonia brunoniana* is the smallest. All have large spade-shaped leaves that produce a dense shade. They commonly grow beside valley streams with nothing beneath them but shade-tolerant ferns and the introduced *ti (Cordyline fruticosa)*. The peculiar wood is very soft and layered like a tightly rolled newspaper. The small flowers develop in clusters at the tips of long stalks.

Fruits and Seeds

Pāpala kēpau fruits ripen in fall and winter. They are 3–7 cm (1–2 in) long *(Pisonia brunoniana* has the smallest) and shaped like miniature grayish green bananas. They occur in loose clusters attached to a slender, multibranched stalk that projects beyond the leaves and is easy to spot. Immature fruits are not sticky, but when mature they exude the resin used by the Hawaiians to catch birds. In nature, however, it is the birds that catch the *pāpala kēpau* fruits. The fruits adhere to the birds on contact, and as they fly away to another site, struggling to divest themselves of their sticky cargo, they serve as a dispersal mechanism for the next generation of the plant. Each fruit contains a single large, elongated seed under a thin layer of flesh.

Germination and Seedling Growth

The traditional and easiest way to propagate *pāpala kēpau* is to simply plant the entire fruit. The fruits remain sticky for days to weeks after removal from the tree but gradually lose their stickiness after planting. Sprouting by this method occurs in four to six weeks. A more rapid method to propagate is by removing the seed from the fruit, which should be done under water as the resin is not sticky when wet. With a small knife or fingernail, make a small incision in the thin wall of the fruit, being careful not to damage the underlining seed. Peel back the fruit wall as you would a banana and remove the soft seed. Plant it directly and horizontally into a tray of new, moist vermiculite. Sprouting takes one to two weeks. Sprouted seeds rapidly develop a root with a thick, white, fibrous mass that remains on or near the surface of the soil while the root's tip grows downward. When removed from the seed bed, this mass binds strongly to the particles of the vermiculite. Meanwhile, the young shoot, with its large, sturdy cotyledons, quickly pulls free of a papery seed coat.

Transfer the young seedlings to individual containers before they develop true leaves. For a short while, two or three weeks, seedlings seem to grow little (above the potting mix). Then, as true leaves appear, the seedling's growth quickens. At two months, same-age seedlings of *Pisonia umbellifera* will be two to three times taller than *Pisonia brunoniana* seedlings. The seedlings use water rapidly, so keep the potting medium moist at all times to prevent wilting. The only pest we have seen on *pāpala kēpau* seedlings is aphids. You can remove these by hand or use a spray of insecticidal soap.

Repotting and Planting Out

Pāpala kēpau is highly tolerant of transplanting. Remarkably, quite large plants, 1–2 m (3–6½ ft) tall, will survive in small 2–3 gallon pots if they are kept well watered. We recommend, however, that you plant your *pāpala kēpau* in the ground when it reaches 20–30 cm (8–12 in) in height. They require moist soil and prefer partial shade but are surprisingly tolerant of hot, dry air and bright sunlight. In the ground, our plants have not grown rapidly—at least not at sea level, where they have remained shrub sized for up to five years. Slugs and snails sometimes attack new foliage. Occasionally, scale insects appear on leaves, but other insect pests are rare, perhaps because of the latexlike sap.

Pilo
(Coprosma rhynchocarpa) — Easy

Habitat

Hawaiian members of the genus *Coprosma* typically grow on mesic to wet mountain slopes and in mountain valleys from about 500 to over 2,000 m (1,600–6,600 ft). *C. rhynchocarpa* is found at intermediate and higher elevations as an understory tree on the island of Hawai'i (Wagner et al. 1990).

Hawaiian Uses

Hawaiians ate the fruits of some *Coprosma* for their laxative properties (Degener 1973). *Coprosma ernodeoides* was used for dyes (Krauss 1993).

This six-month-old *pilo* specimen *(C. rhynchocarpa)* is about 35 cm (1 ft) high and ready for outplanting.

Description

This genus of thirteen species has evolved extensively in Hawai'i (Wagner et al. 1990). *Coprosma rhynchocarpa* is a small tree with thin, delicate branches, dark green, velvety leaves, and bright red berries. Another *pilo, Coprosma ernodeoides,* is a prostrate shrub called *kūkaenēnē* for its small black fruits that resemble the droppings of our state bird, the *nēnē*. Most *Coprosma* have red or orange berries that ripen in fall and winter.

Fruits and Seeds

Pilo seeds are small, 3–4 mm (⅛–³⁄₁₆ in) long, and broadly wedge shaped. One to four seeds commonly occur in the berrylike fruits, whose pulp when ripe is soft and easily removed with your fingers. Seeds may develop a fuzzy coat of mold for several weeks during incubation, but this does not seem to decrease germination.

Germination and Seedling Growth

Pilo seeds take up to four months to sprout. A white root appears at the pointed end of the wedge as the seed coat almost simultaneously sloughs off, revealing a bright green shoot with a hairy reddish stem. Initial growth is variable, but some seedlings may reach a height of 20 cm (8 in) in six months.

Repotting and Planting Out

Our experience with this species in a residential yard near Volcano, Hawai'i, suggests that it is a good understory plant in the company of large *'ōhi'a lehua* and *koa* trees. These *pilo* proved hardy during periods of drought when the young plants were not watered, while, side by side, *kōlea* and *kōpiko* plants died. After five years, these plants have grown to between 2 and 3 m (about 6½–10 ft) tall and begun to flower and fruit. To date, these *pilo* have shown no signs of disease or attack by pests.

Pua kala
(*Argemone glauca*)   **Easy**

Habitat

This is a plant of dry coastal plains and open shrubland slopes from sea level to about 1,900 m (6,200 ft). *Pua kala* occurs in scattered populations on the leeward sides of all the main islands except Kaua'i (Wagner et al. 1990).

Hawaiian Uses

Hawaiians treated toothaches, ulcers, and perhaps arthritis with *pua kala*'s bright yellow sap (Krauss 1993, Stone and Pratt 1994).

Description

With bristling prickles, jagged leaves, and contorted, rambling branches that may extend 2–3 m (6½–10 ft), *pua kala* appears indigenous to certain demonic Tolkien landscapes or perhaps some of the darker Hawaiian mythologies. Yet this plant can be grotesquely beautiful with its overall bluish cast surmounted by prolific displays of large (5–8 cm; 2–3 in), white poppy flowers. The displays can last for weeks in the heat of summer, but the flowers wilt almost instantly when picked.

One of a small number of Hawaiian plants with prickles, spines, or thorns, *pua kala* is a creative solution for keeping pets and children out of the garden. Interestingly, some varieties of *pua kala* no longer have many of their prickles. Unlike the more bristly plants that have prickles covering their leaves, stems, and seed pods, one variety seen on the Big Island (see photo, p. 3) has only prickly leaves. While *pua kala* is reportedly a perennial herb (Wagner et al. 1990), many of our vigorous cultivated plants have died after a single year.

Fruits and Seeds

Pua kala's flowers attract and are effectively pollinated by all types of bees, so good seed is plentiful as early as spring. The many small, spherical, black seeds develop in spiny capsules that grow from the base of each flower. When mature, the capsules become dry and flare open at the end, vaselike. The seeds then spill out onto the ground below as the wind shakes the plant.

With its numerous prickles, *pua kala* is the perfect look-but-don't-touch plant for both gardens and pots. Even if you do manage to pick the flowers without being pricked, they wilt almost instantly.

Germination and Seedling Growth

Pua kala seeds sprout abundantly in the fall. Some of our seed trays have sat for months until the seeds sprouted in November; in others, sown in October and November, the seeds sprouted in two weeks. This suggests that *pua kala* seeds use a timing mechanism to ensure that most of their growth is through the winter and spring—the seasons with the greatest rainfall on the leeward sides of the islands.

Upon germination, *pua kala* sprouts develop long, thin cotyledons and unless you look closely to see their true structure, sprouting seed beds or gardens look as if they are contaminated with grass seedlings. But *pua kala*'s true leaves, with the definitive, deeply incised margins of the genus, develop soon after the cotyledons. Shortly after the second or third true leaves develop, you should transplant the seedlings to small individual pots. Avoid overwatering these young seedlings as they are prone to damping-off diseases.

Repotting and Planting Out

In a pot, *pua kala* becomes root-bound and stunted, but it will still flower after about three months, even if neglected and only several inches tall. To see this plant's greatest growth potential and obtain a succession of flowers for many weeks, you should plant "Hawaiian poppy" in well-drained soil when it is 5–10 cm (2–4 in) tall. (Just don't put it where you will want to walk or work for the next several months.) Plants that grow too quickly tend to fall over, so avoid watering the plant unless you see obvious signs of drought stress such as wilted leaves. After the primary stalks have aged and begun to wither, cut them back and new shoots will often emerge near the base.

Because *pua kala* seeds sprout readily on barren ground (in the fall), you may eventually face a containment problem rather than one of continued propagation. We have had no problems with pests on *pua kala*.

Uhiuhi
(Caesalpinia kavaiensis)   Difficult

Habitat

Nearly extinct in the wild, a few *uhiuhi* still grow in the dry to mesic forests of Waimea Canyon on Kaua'i, O'ahu's Wai'anae Range, and North Kona, Hawai'i (Wagner et al. 1990). It is a federally listed endangered species. On the Big Island it grows in rough, rocky terrain with species such as *kauila (Colubrina oppositifolia), maua (Xylosma hawaiiense), kulu'ī (Nototrichium sandwicense),* and *lama (Diospyros sandwicensis).*

Hawaiian Uses

The Hawaiians had many uses for the heavy, dark wood of *uhiuhi*. It substituted for *kauila* in the making of heavy tools such as 'ō'ō (digging sticks) and weapons such as

After being pollinated, the many deep red flowers of *uhiuhi* develop into leaflike pods, each containing one or two large seeds. This young *uhiuhi* is one of only a few in cultivation; wild trees are rarer still. Fortunately, however, small trees begin producing seeds after only a few years in the ground.

spears, daggers, and clubs (Abbott 1992). They also used it for composite fishhooks, house posts, *kapa* (bark cloth) beaters, and *lā'au kahi wauke,* a board for scraping *wauke* to make *kapa* (Krauss 1993). Finally, the Hawaiians fashioned *uhiuhi* wood into runners for *holua,* sleds used on steep, grassy slopes (Handy and Handy 1972).

Description

With its delicate, light green, compound leaves, *uhiuhi* is reminiscent of the locust trees seen in the continental United States. Mature trees have a thick and rugose bark. The wood is dark (sometimes nearly black), exceedingly hard, and dense; it sinks in seawater. *Uhiuhi* blooms in the winter with large compound clusters of red or orange pealike flowers, each the size of a fingertip.

Today *uhiuhi* is one of Hawai'i's rarest native trees. The three remaining wild populations on Kaua'i, O'ahu, and Hawai'i are all small and continually threatened with extinction. For example, over the last few decades brush fires in the *uhiuhi*'s tinder-dry summer habitat on the Big Island have killed many of the trees. Underneath the few still remaining, the ground is littered with chewed-open pods (almost certainly by rats) with the seeds absent or half eaten. *Uhiuhi* seedlings are all but nonexistent amidst the tall alien grasses. The few seedlings we do come across are likely the product of a researcher's efforts as the small plants are either tagged or the ground surrounding them well tended. On O'ahu, conditions are no better; there, the last *uhiuhi* trees stand with numerous dead limbs. On closer inspection, one can see the cause: the telltale holes of stem-borers.

Fruits and Seeds

Uhiuhi fruits are thin, broad-sided pods containing one or two large seeds that look like brownish lima beans. The fruits and seeds mature from late winter to midsummer.

Germination and Seedling Growth

Older brown seeds that have been dry for some time, perhaps years, take the longest to germinate. The fastest sprouting *uhiuhi* seeds are khaki colored or still retain a faint greenish tint, but they are also full size (about 2 cm; ¾ in long) and very hard. If you are unsure about the seeds' maturity, let the collected pod age for several weeks before extracting the seeds. Modest scarification is important in speeding up germination, which can take as little as a few days to many months. Avoid cutting deeply through the seed coat or shaving it to expose the underlying embryo. Although this will induce immediate water uptake, swelling, and sprouting within days, such sprouts appear more susceptible to rot and may die. A little patience pays off with strong, fast-growing seedlings.

Sprouting is dramatic; the large seed suddenly begins to swell and, over a week or so, reaches two to three times its original size. As it takes on a purplish hue, the seed splits at the narrower end where a stout white root emerges. The shoot arches upward several centimeters before its tip, containing the first vestigial set of compound leaves, pulls free of the split in the seed.

Seedlings grow fast and can be 30 cm (1 ft) tall in a few months. One of our meter-tall saplings flowered in its pot after only one year's growth. We do not advise keeping *uhiuhi* plants in pots for even that long, however, because their long roots can become twisted and hopelessly coiled in the pot. Eventually, as they thicken, the roots begin to strangle the plant. If this happens, you can try to rescue the plant by inducing new roots. Do this by following the same technique previously described for *lama*, which can have the same problem.

Repotting and Planting Out

Uhiuhi tolerates repotting well, but use only rapidly draining potting soil (high in cinder content) because the young plants will not tolerate "wet feet." They respond well to fertilizer (diluted in their water or as controlled-release pellets) and will slow or stop growing when the nutrients are depleted from the potting soil. In young plants, terminal or lateral stems sometimes die back with no apparent cause. In most cases, however, the plant will sprout a new branch lower on the main stem and resume its growth.

Plant *uhiuhi* out at a young age and small size (see below) to avoid root coiling in the pot. After planting out, the young tree may grow quite slowly for two or three months (sometimes longer) before resuming a more rapid growth. Two of our plants, 20–30 cm (8–12 in) tall at the time they were planted in the ground three years ago, are now approximately 2 m (6½ ft) tall and growing rapidly. They began flowering last winter and this winter are covered in flower stalks. Like *kauila, wiliwili,* and other dryland trees, *uhiuhi* grows little during the summer, although we have not seen it stop completely or shed its leaves like *wiliwili.*

Ants of several species, including the fire ant *(Solenopsis geminata)*, will set up nests near *uhiuhi*. There they often foster and protect insect pests such as leafhoppers. These pests, however, do not seem to threaten the life of the plant. In contrast, an unidentified stem-borer can inflict significant and, in the case of a large infestation, potentially fatal damage to *uhiuhi*. Evidence of this borer's attack is wilted leaves, below which you will find a small hole in the stem, 1–2 mm (1/16 in) in diameter, surrounded by "sawdust." Inside the damaged stem is a larva 3–5 mm (1/4 in) long that has effectively eaten all of the tissue within a section of the stem. The only treatment at this point is to remove and destroy the stem along with the insect. Fortunately, based on our experience, this borer seems to be a seasonal pest, with most attacks occurring in the winter and spring. Spraying your *uhiuhi* with a systemic insecticide such as acephate may deter the borer during this period.

Uhiuhi is also susceptible to some pathogenic soil factors, perhaps fungal infections or nematodes, that may be highly localized. *Uhiuhi* planted over several years in one location all succumbed in less than a year, while others from the same parent tree are now three years old and thriving at a different site. A severely affected plant turns pale and wilts. Watering and fertilizer do not correct these symptoms, which are often followed by a general yellowing and dropping of foliage. We suspect that overwatering may even exacerbate this problem. Such a diseased plant dies over a period of weeks to months as its roots seem to disintegrate. We have yet to find an effective treatment for this condition.

Wiliwili
(Erythrina sandwicensis)   Easy

Habitat

This tree thrives in the dry coastal and foothill regions of all the main islands, including Kahoʻolawe and Niʻihau.

Hawaiian Uses

Hawaiians used the soft, light wood for fishnet floats, surfboards, and canoe outriggers (Lamb 1981). The colorful seeds are still strung into leis.

Description

Wiliwili is an extremely hardy tree; wind and drought resistant, it grows up to 15 m (50 ft) tall, but is usually much shorter. The often twisted branches of the tree have inspired several Hawaiian myths (see Neal 1965). While closely related to *Erythrina* species in Tahiti, South America, and the West Indies, the Hawaiian *wiliwili* is

The flowers of Hawai'i's endemic *wiliwili* come in many colors: red, orange, yellow, green, or white. Today it would be hard to find a *wiliwili* large enough to fashion into a traditional Hawaiian surfboard, but the seeds are still common enough to string into leis.

unique in its morphological variation. Individual trees differ in stature, the prevalence of thorns and most notably, flower color. In Hawai'i one can see dwarf *wiliwili*, less than 3 m (10 ft) tall, and giant wiliwili; there are trees with few or no thorns and trees that are quite thorny. *Wiliwili* may have red, orange, yellow, green, or nearly white flowers; red orange is the most common color. (*Wiliwili* species from outside Hawai'i normally have only red or orange flowers.) One popular misunderstanding about our native *wiliwili* is that it is thornless; this is only sometimes true. The trees normally lose their leaves during the dry summer and flower (often still leafless) in late summer to early fall.

Fruits and Seeds

Wiliwili fruits are fuzzy pods containing one to four seeds. The low number of seeds per pod is a good way to distinguish native trees from introduced species. The native *wiliwili* very rarely has pods with more than four seeds, whereas nonnative species typically have many more seeds per pod. Hawaiian *wiliwili* seeds can vary in color from dark red to nearly yellow and are 1–1.5 cm (½ in) long. When mature, the pods split open and, with the help of a strong breeze, the seeds fall to the ground. These seeds are very hard and resist insect attack, perhaps for years. You can often collect seeds in excellent condition by searching under the tree.

Germination and Seedling Growth

The most effective way of speeding up the germination of *wiliwili* seeds is scarification. Do this by sanding a small area of the seed coat. You can also use a sharp knife to scratch the seed coat, but take care because the coat is extremely hard. Place the seeds in a cup of water for several hours. There you will see the seed coats soften and blister and the seeds swell as they take up water. Plant the seeds in any standard pot (although a small, deep pot is best) filled with a porous soil mix. Root growth begins in a few days and the shoot should emerge in one week or less. Initial seedling growth is rapid. After three or four weeks, lightly fertilize the *wiliwili* with any standard all-purpose fertilizer.

Repotting and Planting Out

When your *wiliwili* reaches 30 cm (1 ft) in height, after about two months, either repot it or plant it out in full sun. It is difficult to hasten its growth either with watering or fertilizer after planting out. Instead the young *wiliwili* follows a seasonal growth schedule, growing rapidly during the winter months and little during the summer months. The growth of individual plants is also quite variable; three years after planting out, we have trees of identical age that differ by more than a meter in height.

Wiliwili competes poorly for water and nutrients with common lawn grasses, so keep the area surrounding the tree's base free of grass. Effective methods are using mulch, rocks, or weed-blocking fabric.

During long rainy periods, young *wiliwili* and more mature plants are infected by a white leaf fungus (powdery mildew). Usually the fungus causes little damage other than several dead leaves and disappears without treatment when sunny days return. If it does not, however, treat the plant repeatedly with a standard fungicide spray and move the plant (if possible) to a sunnier location. Leaves are also sometimes attacked by spider mites, which you can eliminate with two or three treatments of insecticidal soap. The small black stinkbug is attracted to *wiliwili* and sometimes damages new stems and leaf petioles.

The most serious pest of *wiliwili* is the Chinese rose beetle, which can defoliate an entire tree in one night. While this will not kill the tree, it is unsightly and repeated attacks can ultimately lead to the plant's demise. (Interestingly and unfortunately, the nonnative *wiliwili* species are far less attractive to the Chinese rose beetle.) Thus far the most effective treatment we have found against the beetle is spraying the tree with acephate (e.g., Isotox at 2 tbs per gallon). The spray will normally protect the plant for two or three weeks, provided there is not a lot of rain during this period (in which case far more frequent spraying is necessary). See the previous section for additional ways to control the Chinese rose beetle.

LANDSCAPE GUIDE

We encounter many people interested in native Hawaiian plants from primarily a landscape point of view. They would like to incorporate natives into their yards but have a specific look in mind (e.g., "I want a groundcover plant for my rock garden."). Because the vast majority of native Hawaiian plants have not experienced horticultural selection, it is difficult to give specific recommendations. For example, wild populations of *naio* can range from a low-growing form found near South Point on Hawai'i, normally less than 30 cm (1 ft) high, to upper forest populations composed of trees over 6 m (20 ft) tall. Even within a single population of wild plants there is often considerable variation in branching, flower size and color, and so on. While botanists delight in this natural variation—the stuff of evolution—most gardeners want to know in advance how the mature plants will look.

The following listing and table summarize much of the information of interest to landscapers for the plants covered in this book. We hope Hawai'i's gardeners will find them helpful when making landscaping decisions.

Groundcovers: 'ākia *(Wikstroemia uva-ursi)*, 'ilie'e, kāwelu, ma'o (some varieties), naio (one variety), nanea, 'ōhelo kai (sparse cover)

Climbing Plants: 'āwikiwiki, maile, nanea

Border Plants: 'a'ali'i, *Achyranthes splendens*, 'āheahea, 'ākia, 'akoko, 'iliahi (coastal variety), 'ilie'e, kāwelu, ko'oko'olau, kōpiko, maiapilo, ma'o, naio, naupaka, 'ōhelo kai, pua kala

Hedges: 'a'ali'i, *Achyranthes splendens*, 'ākia, alahe'e, hō'awa, koki'o ke'oke'o, koki'o 'ula, māmane (open), ma'o, naio, nānū, pāpala kēpau (open)

Ornamental Shrubs and Trees (i.e., shrubs and trees with showy leaves, flowers, or fruits): 'a'ali'i, *Achyranthes splendens*, 'ākia, 'āla'a, alahe'e, hala pepe, hao, hō'awa, 'iliahi, kauila *(Colubrina oppositifolia)*, koai'a, koki'o, koki'o ke'oke'o, koki'o 'ula, kōlea, kōpiko, lama, loulu, maiapilo, māmane, ma'o, maua, nānū, naupaka, 'ohe makai, 'ōhelo kai, pāpala kēpau, pilo, uhiuhi, wiliwili

Shade Trees: 'āla'a, āulu and mānele, hao, 'iliahi (excluding most varieties of *Santalum ellipticum*), kauila *(Alphitonia ponderosa)*, kauila *(Colubrina oppositifolia)*, koa, lama, maua, nānū (old trees), 'ohe makai (seasonal), pāpala kēpau *(Pisonia umbellifera),* wiliwili (seasonal)

Xeriscape Plants: 'a'ali'i, *Achyranthes splendens,* 'āheahea, 'aiea, 'ākia *(Wikstroemia sandwicensis* and *W. uva-ursi),* 'akoko, 'āla'a, alahe'e, āulu and mānele, hala pepe *(Pleomele hawaiiensis* and some varieties of *P. forbesii),* hao, 'iliahi *(Santalum ellipticum* and some varieties of *S. paniculatum),* 'ilie'e (dies back when very dry), kauila *(Alphitonia ponderosa),* kauila *(Colubrina oppositifolia),* kāwelu, koai'a, koki'o, koki'o 'ula (some varieties), ko'oko'olau, lama, maiapilo, māmane, ma'o, maua, naio, *Nama sandwicensis,* nanea, nānū, naupaka, 'ohe makai, 'ōhelo kai, pua kala, uhiuhi, wiliwili

Rock Gardens: 'a'ali'i, *Achyranthes splendens,* 'āheahea, 'ākia, 'akoko, hala pepe, 'iliahi (coastal variety), 'ilie'e, kāwelu, ko'oko'olau, maiapilo, māmane, ma'o, naio, *Nama sandwicensis,* nanea, naupaka, 'ōhelo kai, pua kala

Lei Plants: flowers: 'āwikiwiki, hala pepe, māmane, nānū; fruits: 'a'ali'i; leaves: 'a'ali'i, maile; seeds: āulu and mānele, wiliwili

Container Plants*: 'a'ali'i, *Achyranthes splendens,* 'āheahea, 'ākia, 'akoko, alahe'e, 'āwikiwiki, hala pepe, 'ilie'e, ko'oko'olau, maile, ma'o, naio, *Nama sandwicensis,* nanea, nānū, naupaka, 'ōhelo kai, pāpala kēpau, pua kala

---

* Large container plants often do not survive when planted out because their pot-bound roots are unable to grow properly and serve the plant's needs.

Table 1. Landscape Maintenance Guidelines for Native Hawaiian Plants Featured in This Book

Hawaiian Name	Mature Growth Form	Suggested Cultivation Zones	Watering Requirements One Year After Planting Out	Light Requirements	Pest Problems
ʻaʻaliʻi	shrub or small tree to 8 m (26 ft)	lowland, foothill, mountain	once a month or less during dry months	some direct sunlight each day; does best in full sun	occasional: scale insect, snails and slugs
Achyranthes splendens	shrub	coastal, lowland	once a month during dry months	some direct sunlight each day; does best in full sun	infrequent: scale insect, stem-borer, mealybug
ʻāheahea or ʻāweoweo	shrub	coastal, lowland, foothill, (dry) mountain	once a month or less during dry months	some direct sunlight each day; does best in full sun	occasional: small gray weevil, sucking insects
ʻaiea	small tree to about 10 m (33 ft)	lowland, foothill, (dry) mountain	once a month or less during dry months	some direct sunlight each day; does best in full sun	occasional: stem-chewing pests, transplant shock
ʻākia	shrub to small tree depending on species	coastal, lowland, foothill, mountain depending on species	once or twice a month during dry months	coastal and dryland species prefer full sun; upland species prefer some direct sunlight each day	infrequent: leaf-cutting bee; young plants attacked by snails and slugs
ʻakoko	small to medium sized shrubs depending on species	coastal, lowland	once a month or less during dry months	some direct sunlight each day; does best in full sun	occasional: sucking insects
ʻālaʻa	tree to 15 m (50 ft)	lowland, foothill, mountain	once a month during dry months	some direct sunlight each day; does best in full sun	unknown; have not planted out
alaheʻe	shrub to small tree to 10 m (33 ft)	lowland, foothill, mountain	once a month during dry months	some direct sunlight each day; does best in full sun	occasional: scale insect
āulu and mānele	trees; āulu to 15 m (50 ft), mānele to 25 m (80 ft)	lowland, foothill, mountain	once a month during dry months	some direct sunlight each day; does best in full sun	seasonal: small gray weevil, Chinese rose beetle stemborers
ʻāwikiwiki	climbing vine	lowland, foothill, mountain	once or twice a month during dry months	does best with some direct sunlight each day; can tolerate shade	occasional: stinkbug, scale insect

Table 1 (*continued*)

Hawaiian Name	Mature Growth Form	Suggested Cultivation Zones	Watering Requirements One Year After Planting Out	Light Requirements	Pest Problems
hala pepe	small tree, 5–10 m (17–33 ft)	lowland, foothill, mountain depending on species	once a month during dry months	some direct sunlight each day; does best in full sun	occasional: leaf-spotting disease; infrequent: scale insect, root rot, stem-chewing pests
hao	shrub to small tree, 2–10 m (6–33 ft)	lowland, foothill, mountain	once a month during dry months	some direct sunlight each day; does best in full sun	infrequent: scale insect, leaf-eating larvae
hōʻawa	shrub to small tree, 3–8 m (10–26 ft)	lowland, foothill, mountain depending on species	twice a month or more during dry months	some direct sunlight each day; does best in full sun	occasional: spider mites and sucking insects
ʻiliahi	small shrub to medium-sized tree depending on species	coastal, lowland, foothill, mountain depending on species	once a month during dry months for coastal species, more often for other species	some direct sunlight each day; does best in full sun	occasional: stem-chewing pests when seedling; infrequent: whitefly, scale insect, small gray weevil
ʻilieʻe	sprawling shrub	lowland, foothill, (dry) mountain	once or twice a month during dry months	does best with some direct sunlight each day; can tolerate shade	none encountered
kauila (*Alphitonia ponderosa*)	tree to 25 m (80 ft)	lowland, foothill, mountain	once or twice a month during dry months	some direct sunlight each day; does best in full sun	occasional to frequent: black twig-borer; occasional: whitefly, spider mite
kauila (*Colubrina oppositifolia*)	tree to 13 m (43 ft)	lowland, foothill, (dry) mountain	once a month during dry months	some direct sunlight each day; does best in full sun	occasional: black twig-borer; occasional: small gray weevil
kāwelu	bunch grass about 1 m (3 ft) tall	lowland, foothill, (dry) mountain	once a month during dry months	some direct sunlight each day; does best in full sun	infrequent: leaf-eating pests
koa	tree to 35 m (100 ft) but many varieties much shorter	lowland, foothill, mountain	once or twice a month during dry months	some direct sunlight each day; does best in full sun	numerous: including wilt fungus, stem-borers, leaf-eating insects and sucking insects; damage frequency depends on environment and variety

Table 1 (*continued*)

Hawaiian Name	Mature Growth Form	Suggested Cultivation Zones	Watering Requirements One Year After Planting Out	Light Requirements	Pest Problems
koaiʻa	small tree to 5 m (16 ft)	lowland, foothill, (dry) mountain	once a month during dry months	some direct sunlight each day; does best in full sun	occasional: sucking insects, wilt fungus?
kokiʻo	small tree to 12 m (40 ft) but usually much shorter	lowland, foothill	once a month or less during dry months	some direct sunlight each day; does best in full sun	occasional: sucking insects; infrequent: Chinese rose beetle
kokiʻo keʻokeʻo	small tree to 10 m (33 ft) but can be maintained as a shrub by pruning	lowland, foothill, mountain	twice a month or more during dry months	does best with some direct sunlight each day; can tolerate shade	occasional: sucking insects; infrequent: Chinese rose beetle, bud fly, root fungus or nematode?
kokiʻo ʻula	shrub to small tree, to 8 m (25 ft) but can be maintained shorter by pruning	lowland, foothill, (dry) mountain	once a month during dry months	does best with some direct sunlight each day; can tolerate shade	occasional: sucking insects; infrequent: bud fly
kōlea	small to medium-sized trees depending on species	foothill, mountain	twice a month or more during dry months	does best with some direct sunlight each day; can tolerate shade	unknown
koʻokoʻolau	small herb about 1 m (3 ft) tall	lowland, foothill, mountain	once a month during dry months	some direct sunlight each day; does best in full sun	none encountered
kōpiko	shrubs to medium-sized trees depending on species	lowland, foothill, mountain depending on species	twice a month or more during dry months	does best with some direct sunlight each day; can tolerate shade	infrequent: grasshopper
lama	tree to 10 m (33 ft)	lowland, foothill, mountain	once a month during dry months	some direct sunlight each day; does best in full sun	occasional: stem-tip borer, scale insect, small gray weevil, Chinese rose beetle
loulu	palm trees 5–30 m (17–100 ft) depending on species; most are less than 10 m (33 ft)	lowland, foothill, mountain depending on species	once a month or more during dry months	some direct sunlight each day; does best in full sun	infrequent: stem-borer, whitefly, chewing insects around base

Table 1 (*continued*)

Hawaiian Name	Mature Growth Form	Suggested Cultivation Zones	Watering Requirements One Year After Planting Out	Light Requirements	Pest Problems
maiapilo	low-lying shrub about 1 m (3 ft) tall	coastal, lowland	once a month or less during dry months	some direct sunlight each day; does best in full sun	occasional: leaf-eating pests; infrequent: stem-borer, transplant shock
maile	sprawling shrub about 1 m (3 ft) tall	lowland, foothill, mountain	twice a month or more during dry months	does best with some direct sunlight each day; can tolerate shade	occasional: scale insect, aphid
māmane	shrub to medium-sized tree to 15 m (50 ft) depending on source	lowland, foothill, mountain	once a month or less during dry months	some direct sunlight each day; does best in full sun	infrequent: leaf-eating thrips, Chinese rose beetle
maʻo	shrub 1–2 m (3–6½ ft) tall	coastal, lowland	once a month or less during dry months	some direct sunlight each day; does best in full sun	occasional: cotton aphid, leaf-spotting fungus; infrequent: sucking insects, Chinese rose beetle
maua	small tree to 9 m (30 ft)	lowland, foothill, mountain	once a month during dry months	some direct sunlight each day; does best in full sun	location dependent: black twig-borer; infrequent: Chinese rose beeetle, leaf "burn"
naio	sprawling shrub to small tree depending on source	coastal, lowland, foothill, mountain depending on variety	once a month or less during dry months	some direct sunlight each day; does best in full sun	infrequent: leaf-eating insects, sucking insects, stem-borer
Nama sandwicensis	very small herb	coastal, lowland	once a month or less during dry months	some direct sunlight each day; does best in full sun	infrequent: sucking insects
nanea	crawling vine	coastal, lowland	once a month or less during dry months	some direct sunlight each day; does best in full sun	occasional: bean fly, black stinkbug, leafhopper; infrequent: snails and slugs
nānū	shrub to small tree, to 5 m (17 ft)	lowland, foothill, (dry) mountain	once a month during dry months	some direct sunlight each day; does best in full sun	infrequent: fire ants, scale insect

Table 1 (*continued*)

Hawaiian Name	Mature Growth Form	Suggested Cultivation Zones	Watering Requirements One Year After Planting Out	Light Requirements	Pest Problems
naupaka	sprawling shrub to medium-sized shrub depending on species	coastal, lowland	once a month or less during dry months	some direct sunlight each day; does best in full sun	occasional: spider mite on dwarf naupaka, scale insect
ʻohe makai	tree to 30 m (100 ft) but more often one-third this tall	lowland, foothill, (dry) mountain	once a month or less during dry months	some direct sunlight each day; does best in full sun	infrequent: scale insect
ʻōhelo kai	sprawling shrub less than ½ m (20 in) tall	coastal, lowland	once a month or less during dry months	some direct sunlight each day; does best in full sun	infrequent: sucking insects
pāpala kēpau	small to large trees, to 15 m (50 ft) depending on species	lowland, foothill, mountain	twice a month or more during dry months	does best with some direct sunlight each day; can tolerate shade	occasional: scale insect, aphid, snails and slugs
pilo	prostrate shrub to small tree depending on species	foothill, mountain	once a month or more during dry months	does best with some direct sunlight each day; can tolerate shade	none encountered
pua kala	herb to 2 m (6½ ft) tall	lowland, foothill	once a month or less during dry months	some direct sunlight each day; does best in full sun	none encountered
uhiuhi	shrub to small tree, to 10 m 33 ft)	lowland, foothill, (dry) mountain	once a month or less during dry months	some direct sunlight each day; does best in full sun	seasonal: stem-borer; site specific: soil diseases
wiliwili	small to medium-sized trees to 15m (50 ft) but often much shorter	coastal, lowland, foothill, (dry) mountain	once a month or less during dry months	some direct sunlight each day; does best in full sun	occasional to frequent: Chinese rose beetle, black stinkbug, powdery mildew, spider mite

Appendix 1: Associations, Institutions, and Societies Devoted to Native Hawaiian Plants

On the Big Island

Big Island Native Plant Society (currently inactive). Contact Rex Palmer, David Paul, or Richard M. Waller at P.O. Box 3205, Honoka'a, Hawai'i 96727. (808) 775-0601.

On Kaua'i

National Tropical Botanical Garden. P.O. Box 340, Lāwa'i, Kaua'i, Hawai'i 96765. (808) 332-7324. The garden was formerly a private 100-acre estate started by Queen Emma, wife of Kamehameha IV, and later expanded by the Allertons, a prominent family in Kaua'i during the early twentieth century. It has exceptional plant propagation facilities (for natives and introduced plants) and one of largest germplasm (seeds and other plant parts) collections of Hawaiian natives. There is also a large native plant garden. Volunteers are welcome. Tours are limited; you must make reservations.

On Maui

Native Plant Society (of Maui). P.O. Box 5021, Kahului, Hawai'i 96732. (808) 661-4303. Involved in native plant preservation, including alien pest plant removal.

On O'ahu

Bernice P. Bishop Museum, Department of Natural Sciences/Botany. 1525 Bernice Street, Honolulu. Mailing address: P.O. Box 19000-A, Honolulu, Hawai'i 96817. (808) 848-4175. World's best collection of dried native Hawaiian plants (some of which were collected on Captain Cook's voyages) in the herbarium.

Center for Plant Conservation, Hawai'i Program Office. Located at Lyon Arboretum (3860 Mānoa Road, Honolulu, Hawai'i 96822). (808) 988-1788. The Center's goal is to create a comprehensive national program of plant conservation, research, and education within existing institutions. It helps to maintain rare plants and seed germplasm collections in Hawai'i and elsewhere in the United States.

Earthjustice (formerly Sierra Club Legal Defense Fund), Mid-Pacific Office. 223 South King Street, #400, Honolulu, Hawai'i 96813. (808) 599-2436. Legal environmental group that seeks to change laws and conditions threatening native flora and fauna.

Foster Botanical Garden (part of the Honolulu Botanical Gardens). 50 N. Vineyard Blvd., Honolulu, Hawai'i 96817. (808) 522-7060; twenty-four-hour visitor information at 522-7065. Volunteers are welcome. There is a nice collection of *loulu* palms in the upper terrace and palm garden (refer to the helpful *Self-Guided Tour* book, free at the entrance).

Harold L. Lyon Arboretum. 3860 Mānoa Road, Honolulu, Hawai'i 96822. (808) 988-0456. Part of the University of Hawai'i at Mānoa, this 194-acre garden features tropical plants from all over the world. The arboretum is also a center for research on native and introduced plants. Rare and endangered species (mostly Hawaiian) are propagated and stored in the arboretum's tissue culture laboratory. Native plants are displayed in the Beatrice H. Krauss Ethnobotanical Garden and in the Hawaiian Native Plants section, above/north of the main greenhouse. Volunteers are welcome. Register at Visitor's Center and obtain a self-guided tour map. Public tours are free; a fee is charged for special guided tours.

Hawaiian Botanical Society (sponsored by the Botany Department of the University of Hawai'i at Mānoa). 3190 Maile Way, Honolulu, Hawai'i 96822. (808) 956-8369. Meets the first Monday of every month at 7:30 PM during the school year at St. John Plant Science Laboratory, Room 11. Provides presentations by experts in many areas of botany, including native plants. The society also sponsors field trips to natural areas on O'ahu. Both the presentations and field trips are free and open to the public.

Moanalua Gardens Foundation. 1352 Pineapple Place, Honolulu, Hawai'i 96819. (808) 839-5334. Dedicated to preserving and sharing Hawai'i's unique environmental and cultural resources. Provides guided tours and educational materials. Guided hikes into Kamananui Valley include viewing native plants.

Office of Hawaiian Affairs (OHA). 711 Kapiolani Blvd., Suite #500, Honolulu, Hawai'i 96813. (808) 594-1926. Project led by Roxanne Adams to provide training of gardeners and marketing of native Hawaiian plants for interested Hawaiians as a supplemental business.

The Nature Conservancy of Hawai'i. 1116 Smith Street, Suite 201, Honolulu, Hawai'i 96817. (808) 537-4508. Has offices on Lāna'i, Maui, Moloka'i, and O'ahu. Mission is to preserve rare plants and animals by protecting the lands and water sources they require for survival. Volunteers are welcome. Tours of TNC preserves on several islands are regularly offered.

Waimea Arboretum and Botanical Garden, Waimea Falls Park. 59-864 Kamehameha Highway, Hale'iwa, Hawai'i 96712. (808) 638-8655; Park phone: (808) 638-8511. In a historic Hawaiian valley, the 1,800-acre park features several gardens of native plants as well as many other Hawaiian cultural activities and artifacts. Volunteers are welcome. There is an entrance fee.

Youth for Environmental Service (YES). 1000 Pope Road, Marine Science Building, Room 226, Honolulu, Hawai'i 96822. (808) 957-0423. YES has a nursery at Punahou School and has started planting out native plants at Rocky Hill (on the school grounds) and other locations for restoration and educational purposes. Volunteers are welcome.

Appendix 2: Public and Private Gardens and Sanctuaries Featuring Native Hawaiian Plants

On the Big Island

Amy B. H. Greenwell Ethnobotanical Garden. P.O. Box 1053, Captain Cook. Hawai'i 96704. (808) 323-3318. Located in Captain Cook, 13 miles south of Kailua-Kona, the garden is operated by the Bishop Museum. The garden covers 12 acres, 5 of which are on the remains of a prehistoric Hawaiian agricultural complex. The garden contains an extensive collection of introduced Polynesian and native Hawaiian plants. Free and open to public. Guided tours are available.

Hakalau Forest National Wildlife Refuge (U.S. Fish and Wildlife Service). 154 Waianuinui Avenue, Room 219, Hilo, Hawai'i 96720. (808) 933-6915. This large preserve centered at an elevation of about 1,800 m (6,000 ft) in the upper *koa* and *māmane*-dominated woodland on windward Mauna Kea is inhabited by a number of species of rare plants and native forest birds. Volunteer work is welcomed at Hakalau; it is typically coordinated by organizations such as the Sierra Club and focuses on ecosystem restoration, including planting native trees.

Koai'a Sanctuary (State of Hawai'i). No offices or phone connections at site; call the Department of Land and Natural Resources at (808) 586-8842 for further information. The sanctuary is located in North Kohala, approximately 10 miles north of Waimea on Kohala Mountain Road (Highway 250). Wild and planted native species dominate the site, including of course one of the few remaining populations of *koai'a*.

Koki'o Sanctuary (State of Hawai'i). No offices or phone connections at site; call the Department of Land and Natural Resources at (808) 586-8842 for further information and entry permission. The sanctuary is located in North Kona, approximately 15 miles north of Kailua-Kona on the Hawai'i Belt Road (Highway 190). A small fenced enclosure contains many rare, planted native trees and shrubs, including *hala pepe, kauila, koki'o,* and *uhiuhi*.

Manuka State Park (State of Hawai'i). No offices or phone connections at site; call the Department of Land and Natural Resources at (808) 586-8842 for further information. Located 19.3 miles west of Nā'ālehu, off Highway 11, the park displays some rare native plants of the region such as *koki'o*.

National Tropical Botanical Garden Sanctuary. No offices or phone connections at site; call the National Tropical Botanical Garden at (808) 332-7361 for further information and entry permission. Located approximately 10 miles north of Kailua-Kona on the Hawai'i Belt Road (Highway 190). The site contains many rare and endangered wild plants as well as many newly planted native species.

Sadie Seymour Botanical Gardens. 76-6280 Kuakini Highway, Kailua-Kona, Hawai'i 96740. (808) 329-7286. These gardens, part of the Kona Outdoor Circle Education Center, feature many cultivated plants of Hawai'i, including some native species.

On Kaua'i

Kīlauea Point National Wildlife Refuge (U.S. Fish and Wildlife Service) P.O. Box 87, Kīlauea, Hawai'i 96754 (808) 828-1413. Scenic hills and cliffs overlooking the sea are nesting sites for a variety of seabirds amidst native coastal vegetation, including the rare lobelioid, *Brighamia insignis*.

Kōke'e Museum and Park (State of Hawai'i). P.O. Box 100, Kekaha, Hawai'i 96752. (808) 335-9975. There is an excellent exhibit of native woods in the museum. A self-guided loop trail behind the museum features several labeled native plants, not to mention the miles of trails in the park providing access to many native plants in their natural habitat.

Limahuli Gardens (part of the National Tropical Botanical Gardens). Contact the Administrative Office at P.O. Box 340, Lāwa'i, Hawai'i 96765. (808) 332-7324. Located in Hā'ena half a mile past the 9-mile marker on Kūhiō Highway (Highway 560). This 15-acre garden and 985-acre forest preserve features introduced Polynesian and native Hawaiian species. Both guided and self-guided tours are available for a fee. Reservations are required at (808) 826-1053.

National Tropical Botanical Garden. *See* Appendix 1.

On Maui

Haleakalā National Park, Visitor's Center. P.O. Box 369, Makawao, Hawai'i 96768. (808) 572-4400. Features many native plants of the park in surrounding gardens.

Kahanu Garden (part of the National Tropical Botanical Gardens). 'Ālau Place, Hāna, Hawai'i 96713. (808) 332-7361. Site of Pi'ilanihale Heiau, this 123-acre garden features plants from the Pacific Islands. It has a large collection of breadfruit cultivars and a recently developed coastal garden featuring native plants of Maui. Guided tours are available at (808) 248-8912. There is an entrance fee.

Kahului Public Library Courtyard Garden. 90 School Street, Kahului, Hawai'i 96732. (808) 873-3097. Featuring about twenty species of native plants, the garden is designed and maintained by the Native Hawaiian Plant Society. The garden can be viewed only from inside the library.

Maui Botanical Gardens. Kanaloa Avenue, Wailuku, Hawai'i 96793. (808) 243-7397. Sponsored by the Maui County Horticulture and Beautification Department, this is a 7-acre park devoted to native Hawaiian and introduced Polynesian plants.

Wailea Point Seawalk. Located on the south shore in a private resort community, public access is between the Kea Lani and Four Seasons hotels. This half-mile stretch along the ocean features planted native coastal plants. The walk is free and open to public.

On O'ahu

Bernice P. Bishop Museum. 1525 Bernice Street, Honolulu. Mailing address: P.O. Box 19000-A, Honolulu, Hawai'i 96817. (808) 847-3511. A number of sites on the grounds of the museum have native plants.

Cultural Learning Center at Ka'ala. P.O. Box 630, Wai'anae, Hawai'i 96792. (808) 696-4954. Involved in native plant propagation and restoration, along with Hawaiian cultural, economic, and language activities.

Department of Land and Natural Resources (State of Hawai'i). 1151 Punchbowl, Honolulu, Hawai'i 96813. (808) 586-8842. At the headquarters is a native coastal plant and *loulu* palms garden.

Foster Botanical Garden. See Appendix 1.

Hālawa Xeriscape Garden. 99-1268 Iwaena Street, 'Aiea, Hawai'i 96701. (808) 527-6113 or 527-6126. This 3-acre demonstration garden of the Honolulu Board of Water Supply features water-conserving plants, including a number of native plants. Open hours are limited. A self-guided tour brochure is available at the office. Free to the public.

Harold L. Lyon Arboretum. See Appendix 1.

Ho'omaluhia Botanical Garden (part of the Honolulu Botanical Gardens). 45-680 Luluku Road, Kāne'ohe, Hawai'i 96744. (808) 233-7323-4. Located on the windward (cooler and moister side) of O'ahu, at the base of the Ko'olau Range. This 400-acre garden contains a large number of tropical plants from all over the world, plus a reservoir. Native Hawaiian plants are conveniently located in the Kahua Lehua section and are labeled. Check the visitor center for guided hikes. Free to the public.

Ka Papa Lo'i O Kānewai. Located on Dole Street, at the 'Ewa (west) side of the Hawaiian Studies Center on the Mānoa Campus. No offices or phone connections at site; call the Hawaiian Studies Center at (808) 973-0989 for further information. Built around the site of several old irrigation channels, the gardens have a number of native and introduced Polynesian plants.

Koko Crater Botanical Garden (part of the Honolulu Botanical Gardens). No offices or phone connections at site; call Foster Garden at (808) 522-7060 for further information. A 200-acre garden on the inner slopes and basin of Koko Crater. There is an excellent collection of native plants; most are dryland species and are labeled. There is also a grove of large native *wiliwili* trees. Free to the public.

Leeward Community College. 94-045 Ala 'Ike, Pearl City, Hawai'i 96782. (808) 455-0285. Several sites on campus are planted with coastal, dryland, or mesic forest native species.

Lili'uokalani Botanical Garden (part of the Honolulu Botanical Gardens). No offices or phone connections at site; call Foster Garden at (808) 522-7060 for further information. The 7½-acre garden is located between N. Kuakini and School Streets in Honolulu, following the riverbed of Nu'uanu Stream. It was once part of Queen Lili'oukalani's picnic grounds. There is a nice collection of dryland, coastal, and mesic native plants; most are labeled. Free to the public.

Mau'umae Nature Park, Kaimukī. Sponsored by Mau'umae Nature Park Committee and Department of Parks and Recreation, City and County of Honolulu. (808) 973-7250.

The base park is located at 16th Avenue and Claudine Street, Kaimukī. A trail begins at the back of park and goes upward and across the street to a loop trail up the side of the valley. The hike takes about one hour. The park has 33 acres and volunteers are actively outplanting dryland natives and some introduced Polynesian plants in the mostly nonnative-dominated drylands.

Moanalua Gardens. *See* Appendix 1.

Queen Kapiʻolani Garden (part of the Honolulu Botanical Gardens). No offices or phone connections at site; call Foster Garden at (808) 522-7060 for further information. Located at the corner of Lēʻahi and Monsarrat Avenues in Waikīkī, this garden features an assortment of native dryland plants.

Urban Garden Center. 962 Second Street, Pearl City, Hawaiʻi 96782. (808) 453-6050. This 30-acre garden features many edible and ornamental plants, including some native plants. Open hours are limited. Check the office for tours. Free to the public.

Wahiawā Botanical Garden (part of the Honolulu Botanical Gardens). 1396 California Avenue, Wahiawā, Hawaiʻi 96786. (808) 621-7321. In central Oʻahu and located at a slightly higher elevation than the other Oʻahu botanical gardens, this 27-acre garden is mainly in a forested ravine. It has an excellent native plant collection in the southern end of the riverbed valley. Most are labeled, and they include both mesic and dryland forest plants. Free to the public.

Waikikī Aquarium. 2777 Kalākaua Avenue, Honolulu, Hawaiʻi 96815. (808) 923-9741. The aquarium has a native coastal garden next to its outdoor reef display. There is an entrance fee.

Waimea Arboretum and Botanical Garden. *See* Appendix 1.

Glossary

Achene: A small dry fruit of certain plants characterized by a closed husk.

Adaptive Radiation: A burst of evolution producing several to many new species from a common ancestor. Commonly exhibited by plants and animals that spread among different islands of an archipelago.

Annual Plant: Grows to maturity, produces seeds, and dies within one year.

Arborescence: The evolution of treelike characteristics (extensive woody tissue and large stature) in plants whose ancestors were typically small and soft stemmed.

Axil: The upper angular space formed where two stems branch or where a leaf adjoins a stem.

Birdlime: Term used by early British explorers to refer to sticky resinous sap or secretions of several kinds of plants. In Hawaiʻi, feather hunters smeared this substance on platforms of sticks or defoliated tree branches in order to trap small birds. The native *pāpala kēpau* tree and the introduced breadfruit tree were sources of birdlime.

Calcareous: Having layers or deposits of calcium carbonate, or limestone.

Cotyledon: Structure in a seed that conveys stored nutrients to a sprouting seedling. Twin cotyledons characterize dicot plants such as most kinds of trees, shrubs, and vines. These embryonic structures commonly unfold as the seed spouts, becoming the first set of "seed leaves." Monocot plants such as grasses, lilies, and palms retain the cotyledon inside the seed as the embryo emerges.

Deciduous: Referring to a plant that loses its leaves seasonally. Some dry-forest tropical trees are summer-deciduous, avoiding excessive water loss in the driest months.

Dehisce: To split open, the action of a seed capsule releasing seeds.

Dioecious: A species whose individuals are of separate sexes (male and female); common in animals, rarer in plants.

Endemic: A species that is native by virtue of having evolved in a particular geographic location and found naturally only in that location.

Endocarp: The innermost layer of the fruit that directly surrounds the seed.

Epicotyl: Tissue in a seed that gives rise above the cotyledon to the embryonic stem, leaves, and flowers.

Exocarp: The outer layer (skin or husk) of a fruit or seed-containing structure.

Germplasm: The embryonic tissue of a plant's seed whose cells will divide to give rise to the seedling. Also, active embryonic tissue at the tips of stems and roots that persists through the life of the plant.

Hilum: A small region of a seed at which the embryonic tissue is attached to the endocarp.

Hypocotyl: The embryonic stem below the cotyledon ending at the root.

Indigenous: A species that is native in a given region by virtue of having spread through the region on its own, but whose site of evolutionary origin is unspecified.

Lignified: Transformed from soft tissue into a harder, woody texture. Containing lignin, a chemical component of wood.

Mesic: Climate of moderate rainfall—between dry and wet extremes for a given region.

Mesocarp: An inner protective layer in a seed-containing capsule, or fruit. The mesocarp lies between the exocarp and endocarp, or seed coat.

Monoecious: A species whose individuals combine sexes, having both male and female reproductive organs.

Perennial Plant: Typically a large species that matures slowly and lives for many years.

Petiole: The stalk or stemlike portion of a leaf that attaches it to the stem.

Phyllodes: Leaflike structures in some plants that develop from a flattening of the petioles, or leaf stalks.

Scarification: Process of eroding or removing the surface coating of seeds that often inhibits or slows sprouting. Artificial physical scarification typically employs a knife, file, or sandpaper; chemical scarification may involve acids or hot water treatment.

Stasis: A state of static balance or equilibrium; stagnation.

Succulent: Plant, or condition of a plant, having a high water content, typically in leaves that appear swollen with the stored water. The succulent condition is an adaptation to dry habitats.

Taxonomy: The scientific classification of living organisms.

Transpiration: The evaporative loss of water from a plant's leaves.

Utricle: A small husklike capsule or fruit enclosing a single seed. When ripe, the capsule typically opens spontaneously at one end to expose the seed.

Vermiculite: An artificially prepared soil component or additive manufactured by heating natural mineral products to very high temperatures. Available in all garden supply stores, it has excellent properties as a growth medium for seedlings.

Xeriscape: A landscape planted with drought-resistant plants.

Literature Cited

Abbott, I. A. 1992. *La'au Hawai'i: Traditional Hawaiian Uses of Plants.* Bishop Museum Press, Honolulu.

Alexander, M. C. 1934. *William Patterson Alexander in Kentucky, the Marquesas, Hawaii.* Privately printed, Honolulu.

Anderson, R. C., and D. E. Gardner. 1998. "Investigations of Koa *(Acacia koa)* Decline in Hawaiian Forests." Abstract of poster presented at Hawai'i Conservation Conference, Honolulu.

Arrigoni, E. 1978. *A Nature Walk to Ka'ena Point.* Topgallant, Honolulu.

Bates, G. W. ("A Haole"). 1854. *Sandwich Isle Notes.* Harper and Brothers, New York.

Beardsley, J. W. 1979. "New Immigrant Insects in Hawaii: 1962 through 1978." *Proc. Haw'n Entomol. Soc.* 13(1): 35–44.

Bebb, C. A. S. 1998. "Tree Planting Myths Dispelled." *Hawai'i Horticulture* 1(2): 11–13.

Bornhorst, H. L. 1996. *Growing Native Hawaiian Plants: A How-to Guide for the Gardener.* Bess Press, Honolulu.

Bornhorst, H. L., and F. Rauch. 1994. *Native Hawaiian Plants for Landscaping, Conservation, and Reforestation.* Research Extension Series 142, Hawai'i Institute of Tropical Agriculture and Human Resources, University of Hawai'i, Honolulu.

Carlquist, S. 1980. *Hawai'i: A Natural History.* 2nd ed. Pacific Tropical Botanical Garden, Lāwa'i, Kaua'i.

Culliney, J. L. 1988. *Islands in a Far Sea: Nature and Man in Hawaii.* Sierra Club Books, San Francisco.

Degener, O. 1973. *Plants of Hawaii National Parks: Illustrative of Plants and Customs of the South Seas.* Privately printed.

Degener, O., and I. Degener. 1957. *Flora Hawaiiensis: Family 305, Rauvolfia sandwicensis.* Published privately, 2 pp.

Fleming, G., and C. E. Hess. 1965. "The Isolation of a Damping-Off Inhibitor from Sphagnum Moss." *Proc. Inter. Plant Prop. Soc.* 14: 153–154.

Funasaki, G. Y., P. Lai, L. M. Nakamura, J. W. Beardsley, and A. K. Ota. 1988. "A Review of Biological Control Introductions in Hawaii: 1980 to 1985." *Proc. Haw'n Entomol. Soc.* 28: 105–159.

Ganders, F. R., and K. M. Nagata. 1984. "The Role of Hybridization in the Evolution of *Bidens* on the Hawaiian Islands." In W. F. Grant, ed., *Biosystematics.* Academic Press Canada, Ontario, pp. 179–194.

Gressitt, J. L., and C. M. Yoshimoto. 1964. "Dispersal of Animals in the Pacific." In J. L. Gressitt, ed., *Pacific Basin Biogeography: A Symposium.* Bishop Museum Press, Honolulu, pp. 283–292.

Handy, E. S. C., and E. G. Handy. 1972. *Native Planters in Old Hawaii: Their Life, Lore, and Environment.* B. P. Bishop Museum Bulletin 233. Bishop Museum Press, Honolulu.

Hawaii Audubon Society. 1981. *Hawaii's Birds.* Hawaii Audubon Society, Honolulu, pp. 77.

Highlights, Nitrogen Fixing Tree Association. Morrilton, Arkansas.

Hsiao, A. I., and W. A. Quick. 1984. "Actions of Sodium Hypochlorite and Hydrogen Peroxide on Seed Dormancy and Germination of Wild Oats, *Avena fatua* L." *Weed Research* 24: 411–419.

Kirch, P. V. 1985. *Feathered Gods and Fishhooks: An Introduction to Hawaiian Archeology and Prehistory.* University of Hawai'i Press, Honolulu.

Krauss, B. H. 1993. *Plants in Hawaiian Culture.* University of Hawai'i Press, Honolulu.

Lamb, S. H. 1981. *Native Trees and Shrubs of the Hawaiian Islands.* The Sunstone Press, Santa Fe, New Mexico.

Lewin, R. 1992. *Complexity: Life at the Edge of Chaos.* Macmillan, New York.

Lydgate, J. M. 1883. "Hawaiian Woods and Forest Trees." *Thrum's Hawaiian Almanac and Annual for 1883,* pp. 33–35.

Mikkelson, D. S., and M. N. Sinah. 1961. "Germination Inhibition in *Oryza sativa* and Control by Preplanting Soaking Treatments." *Crop Science* 1: 332–335.

Nagata, K. 1992. *How to Plant a Native Hawaiian Garden.* State of Hawai'i, Office of Environmental Quality Control, Honolulu.

National Tropical Botanical Garden. 1992. "Plant Information Sheets on Native Plants of Hawaii: Ma'o." National Tropical Botanical Garden, Lāwa'i, Kaua'i.

Neal, M. C. 1965. *In Gardens of Hawai'i.* Bishop Museum Special. Publication 40. Bishop Museum Press, Honolulu.

Olson, S. L., and H. F. James. 1982. "Fossil Birds from the Hawaiian Islands: Evidence for Wholesale Extinction by Man before Western Contact." *Science* 217: 633–635.

Price, A. G., ed. 1971. *The Explorations of Captain James Cook in the Pacific as Told by Selections of His Own Journals, 1768–1779.* Dover, New York, p. 216.

Proctor, V. W. 1968. "Long Distance Dispersal of Seeds by Retention in Digestive Tract of Birds." *Science* 160: 321–322.

Rock, J. F. 1974. *The Indigenous Trees of the Hawaiian Islands.* Pacific Tropical Botanical Garden, Lāwa'i, Kaua'i.

Rosa, K.D. 1994. "Acacia Koa—Hawai'i's most valued native tree." NFT Highlights, Nitrogen Fixing Tree Association. Morrilton, Arkansas.

Smith, C. W. 1985. "Impact of Alien Plants on Hawai'i's Native Biota." In C. P. Stone and J. M. Scott, eds., *Hawai'i's Terrestrial Ecosystems: Preservation and Management.* Cooperative National Park Resources Studies Unit, University of Hawai'i, Honolulu, pp. 180–243.

Sohmer, S. H., and R. Gustafson. 1987. *Plants and Flowers of Hawai'i.* University of Hawai'i Press, Honolulu.

Stone, C. P., and L. W. Pratt. 1994. *Hawaii's Plants and Animals: Biological Sketches of Hawai'i Volcanoes National Park.* Hawai'i Natural History Association (distributed by University of Hawai'i Press, Honolulu).

Tenbrink, V. L., and A. H. Hara. 1994. "*Xylosandus compactus* (Eichoff) (Black Twigborer)." Crop Knowledge Master. Internet URL: http://www.extendo.hawaii.edu/kbase/crop/Type/xylosand.htm

Wagner, W. L., D. R. Herbst, and S. H. Sohmer. 1990. *Manual of the Flowering Plants of Hawai'i*. Vols. 1 and 2. Bishop Museum Special Publication 83. University of Hawai'i Press and Bishop Museum Press, Honolulu.

Whistler, W. A. 1980. *Coastal Flowers of the Tropical Pacific: A Guide to Widespread Seashore Plants of the Pacific Islands*. Pacific Tropical Botanical Garden, Lāwa'i, Kaua'i.

Williams, F. X. 1931. *Handbook of the Insects and Other Invertebrates of Hawaiian Sugarcane Fields*. Hawaiian Sugar Planters Association. Advertiser Publishing Co., Honolulu, pp. 194–198.

Zimmerman, E. C. 1958. *Insects of Hawaii*. Vol. 8. *Lepidoptera: Pyraloidea*. University of Hawai'i Press, Honolulu.

Index

Major discussions are in **boldface**.

'a'ali'i, 23, 26, 32, **35–37**, 41, 100, 112
Acacia koa. See *koa*
Acacia koaia. See *koai'a*
acephate, 24, 26, 28–29, 37–38, 41, 48, 50, 52, 55, 64, 68, 77, 103, 105, 107, 112, 114, 128, 139, 141
Achatina fulica. See African snail
achene, 2, 95
Achyranthes splendens, 25–26, 31, **37–39**
adaptations, 2, 7–8, 21
adaptive radiation, 46, 62
Adoretus sinicus. See Chinese rose beetle
African snail, 98
'ahakea, 92
'āheahea, 2, **39–41**
'aiea, 2, 4, 19, 20, **41–43**
'ākia, 2, 16, 20, **43–46**, 62, 67, 98, 121
'akoko, 2, 40, **46–48**, 62
'āla'a, 12, 16, **48–50**
'ala'ala wai nui, 105
alahe'e, 26, **50–53**, 98, 127
'alalā, 62
alien species, x, 4–6, 14–16, 23–33; insects, 15–16, 24–31; mites, 29; nematodes, 31–32; plants— competition with natives, 4–6; rodents, 3, 5, 105; snails and slugs, 23; ungulates, 2–4. *See also accounts of impacts on specific native plants (Section 3)*
Alphitonia ponderosa. See *kauila*
Alticinae. See flea beetle
Alyxia oliviformis. See *maile*
'amakihi, 108
ant, 26–29, 31, 37–38, 48, 52, 61, 83, 86, 107, 112, 119, 128; fire, 123, 139
'apapane, 108
aphid, 26–28, 31, 41, 43, 63–64, 80, 85–86, 88–89, 91, 97, 105, 111–112, 117, 119, 123, 128, 130, 132; cotton, 111–112
Aphis gossypii. See aphid
arborescence, 2, 39–40, 46–47
Argemone glauca. See *pua kala*
Artageia rapae. See cabbage butterfly
āulu, 4, 16, 24, 25, 30, 32, **53–55**, 127
'āweoweo. See *'āheahea*
'āwikiwiki, 28, **55–57**, 80

bacteria, 12, 14–15, 32, 50, 54, 58, 71, 74, 80, 109, 116
bark cloth. See *kapa*
bastard sandalwood. See *naio*
bean fly, 30, 120
bee, leaf-cutting, 46
Bidens amplectens. See *ko'oko'olau*
Bidens pilosa, 94
Bidens torta. See *ko'oko'olau*
Bishop Estate, 4
biodiversity, x, 3
birdlime tree, 48, 131
bleach treatment, 14–15, 36, 42, 45, 63, 69, 74, 80, 83, 123, 127, 129
Bobea sp. See *'ahakea*
Broussonetia papyrifera. See *wauke*
bulbul, 104

cabbage butterfly, 26
Caesalpinia kavaiensis. See *uhiuhi*
Canavalia galeata. See *'āwikiwiki*
Canavalia hawaiiensis. See *'āwikiwiki*
Canavalia kauaiensis. See *'āwikiwiki*
canoe, 41, 46, 61, 77–79, 81, 92, 139
Capparis sandwichiana. See *maiapilo*
Capparis spinosa, 103
Captain Cook, 3, 14, 23
Captan, 66
caryopses, 76
cattle ranching, 4
Chamaesyce celastroides. See *'akoko*
Chamaesyce degeneri. See *'akoko*
Chamaesyce hypericifolia, 67
Chamaesyce skottsbergii. See *'akoko*
chaos theory, 10
chemical inhibitors to sprouting, 13
Chenopodium album. See lambs quarters
Chenopodium oahuense. See *'āheahea*
Chinese rose beetle, 23–25, 55, 77, 80–81, 83, 86, 89, 98, 101, 109, 112, 114, 117, 123, 128, 141
Chloris barbata, 126
Coccus viridis. See scale insect
Colletotrichum. See fungus, leaf spotting
Colubrina oppositifolia. See *kauila*
controlled experiments, 34

Index

Coprosma ernodeoides. See *pilo*
Coprosma rhynchocarpa. See *pilo*
Coptosoma xanthogramma. See stinkbug, black
cordage plant, 45, 86
Cordyline fruticosa. See ti
Corvus tropicus. See *'alalā*
cotton aphid. See aphid
Criley, Richard, 52
cuttings, 7, 9, 16–17, 48, 51–52, 88, 90–91, 116–117, 119, 123; collecting, 7, 16–17; misting technique, 17, 88; rooting, 16–17, 48, 91

damping-off, 20, 32, 46, 104
deforestation, 3–5
Diazinon, 22, 27, 31, 38, 43, 59, 61, 83, 103
dioecious, 112–113
Diospyros hillebrandii. See *lama*
Diospyros sandwicensis. See *lama*
diseases, 12, 14–15, 17, 20, 23–24, 32–33; bacterial, 32, 50, 54, 71, 74; damping-off, 20, 104, 136; fungal, 32, 50, 66, 71–72, 74, 80–81, 83, 89, 100, 104, 111, 134, 141; of leaves, 32, 59, 75, 111, 141; resistance to, 45–46, 61, 71, 108; of roots, 139; of seeds, 50, 58, 74, 83, 92, 134; of stems, 74–75, 100; treatment and prevention, 15, 18, 32, 71–72, 75, 79, 88, 96, 116; viral, 32. See also *further accounts relating to specific plants (Section 3)*
Dodonaea viscosa. See *'a'ali'i*
dracaena, 58

Empoasca sp. See leafhopper, green
Endangered Species Act, 5
endocarp, 92–93, 102, 115
Eragrostis variablis. See *kāwelu*
Erythrina sandwicensis. See *wiliwili*
ethics of growing native plants, 6–11
evolution, ix, 1–2, 39–40, 45–47, 62, 84
'Ewa hinahina. See *Achyranthes splendens*
exocarp, 102
extinction of native flora, ix-x, 2–6

federal laws, 9
fertilizing, 20, 22; new transplants, 19, 22; potted plants, 20; young plants, 16, 19–20. See also *accounts relating to specific plants (Section 3)*
fire, 3–6, 41, 73, 84, 101, 110, 137
flea beetle, 26
Flueggea neowawraea. See *mēhamehame*
foliage-consuming pests. See pests
founders of Hawaiian flora, 1–2
fountain grass, 73
fruits: artificial ripening of, 12, 49, 58, 63, 73–74, 104; collecting, 12, 76; extracting seeds from, 12, 14, 45, 69, 71, 92, 102, 129, 132. See also *descriptions in accounts of individual species (Section 3)*
fungus, 17, 30, 32, 46, 50, 66, 80–81, 83, 89, 100, 104, 111, 117, 141; leaf spotting, 111

Gardenia brighamii. See *nānū*
Gardenia mannii. See *nānū*
Gardenia remyi. See *nānū*
genetic diversity, 7–8, 38, 76
gibberellic acid, 65–66, 125
gnats, 15
Gossypium, 84. See also *ma'o*
Gossypium tomentosum. See *ma'o*
grasses, alien impacts, 4–6, 41, 73; beneficial effects, 25; native species, 75–77
grasshopper, 26, 77, 98, 103, 117
green moss, 15, 40, 47, 59, 63, 77, 88, 95, 99, 116, 123, 129
growth rates. See *individual species accounts (Section 3)*

hala pepe, 4, 12, 15, 19–20, 22, 32, **57–59**, 76, 121
hālau hula, 58, 98, 105
hale, 35, 70, 79, 91, 98
Hancock, Judy and Will, xi, 82
hao, 14, **59–61**, 127
Hawaiian beach pea. See *nanea*
Hawaiian Botanical Society, 6
Hawaiian cotton. See *ma'o*
Hawaiian crow. See *'alalā*
Hawaiian poppy. See *pua kala*
Hawaiian violet, 40
Heteropogon contortus. See *pili* grass
Hibiscus arnottianus. See *koki'o ke'oke'o*
Hibiscus brackenridgei. See *ma'o hau hele*
Hibiscus clayi. See *koki'o ula*
Hibiscus kokio. See *koki'o ula*
hinahina, 16–17
hō'awa, 30, 32, **61–64**
hopper burn, 29

'iliahi, 4, 14, 32, **64–68**, 105, 114, 121
'ilie'e, 13, 16, **68–70**
'ilima, 41, 53, 110
insecticidal soap, 26, 28–29, 37–38, 41, 43, 48, 52, 55, 61, 64, 68, 72, 80, 83, 85–86, 88, 91, 97, 103, 107, 111–112, 119, 123, 126, 128, 130, 132, 141
Integrated Pest Management Program, 24

kāhili, 72, 77
kapa, 35, 64, 70, 72, 91, 96, 98, 105, 110, 121, 137
Kapo, 58
kauila, 4–5, 12, 16, 20, 25, 30, 32, 100, 127, 136, 138; *Alphitonia*, 29, **70–72**; *Colubrina*, 31, **72–75**, 136
kāwelu, **75–77**
koa, 4, 24–25, 30, 67, **77–81**, 82–83, 107–109, 134
koai'a, xi, **81–83**
koki'o, 29, **83–86**
koki'o ke'oke'o, 23, **89–89**
koki'o 'ula, 6–7, 16–17, 88, **89–91**
Kokia drynarioides. See *koki'o*
kōlea, 14, 16, 43, 62, **91–93**, 96, 100, 134

Index

koʻokoʻolau, 2, 32, 62, **93–96**
kōpīko, 16, 43, 62, 93, **96–98**, 100, 134
kūkaenēnē, 134
kuluʻī, 2, 7, 16, 40, 136

lady beetle, 27–28, 121, 130
Laka, 98, 105
lama, 4, 13, 16, 30, 32, 43, 50, 93, 96–97, **98–101**, 105, 112, 121, 136, 138
lambs quarters, 39
leafhopper, 28–29, 83, 86, 139; green, 29; two-spotted, 29
lobeliods, 40
loulu, 62, 67, **101–103**
Lycium sandwicense. See *ʻōhelo kai*

māhapilo, 12–13, 26, 43, **103–105**
maile, 16, **105–107**, 109
Maile sisters, 105
malathion, 28, 114
māmane, 16, **107–109**
mānele, 53–55
maʻo, 29, 32, 41, 53, 84, **109–112**
maʻo hau hele, 6, 9, 24, 86, 88
maua, 19, 30, **112–114**, 136
mealybug, 26–28, 31, 41, 48, 52, 63, 80, 83, 86, 119, 130; cottony, 38
mēhamehame, 30, 112
Melodogyne sp. See nematode, root-knot
mesocarp, 102
millipede, 22, 30, 43
mites, 29, 43, 63, 72, 83, 85, 123, 126, 128, 141
monoecious, 113
Montgomery, Steve, xi, 51
mulch, 40
Myllocerus sp. See weevil, small gray
Myoporum sandwicense. See *naio*
Myrsine lanaiensis. See *kōlea*
Myrsine lessertiana. See *kōlea*

naʻenaʻe, 40
naio, 7–8, 16–17, 26, 53, **114–117**, 118
Nama sandwicensis, **117–119**
nanea, 25, 29–30, **119–121**
nānū, 6, 13, 29, **121–123**
native insects, 23, 79
Native Plant Society of Maui, 6
naupaka, 16, 29, **124–126**
nehe, 16–17
nematode, 31–32, 89, 111, 117, 139; root-knot, 31–32
nematode-resistant natives, 32
Nezara viridula. See stinkbug, green
nitrogen-fixation, 80, 109
Nothocestrum breviflorum. See *ʻaiea*
Nototrichium sandwicense. See *kuluʻī*

ʻōhai, 6, 24–26, 28, 121
ʻohe makai, 4, 22, 98, **126–128**

ōhelo, 44, 129
ʻōhelo kai, **128–130**
ʻōhiʻa lehua, 4, 7, 67, 93, 96, 105, 107, 134
Oidium sp. See powdery mildew
olopua, 96
ōʻō (digging stick), 48, 70, 108, 136
Ophiomyia phaseoli. See bean fly
Orneodes objurgatella, 51

palila, 108
pāpala kēpau, 12, 30, 50, **130–132**
pāʻūohiʻiaka, 16
Pele, 125
Pennisetum setaceum. See fountain grass
Pests, x, 3, 5–6, 7–8, 14, 15–16, 22, 23–32; of foliage, 24–26, 29; of roots, 27, 31–32, 59; of seeds, 15; of stems, 28, 29–31, 43, 103; treatments and prevention, 22, 23–32, 43. See also under individual accounts of native plant species (Section 3)
pig, 3–4, 11, 68
pili grass, 75
pilo, 62, **133–134**
Pisonia brunoniana. See *pāpala kēpau*
Pisonia grandis, 131
Pisonia sandwicensis. See *pāpala kēpau*
Pisonia umbellifera. See *pāpala kēpau*
Pittosporum confertiflorum. See *hōʻawa*
Pittosporum flocculosum. See *hōʻawa*
Pittosporum hawaiiense. See *hōʻawa*
Pittosporum hosmeri. See *hōʻawa*
Planchonella, 49
planting out, 20–23; appropriate habitat, 20–21, 117; drainage, 21–22; hole size, 21; restrictions, 7–8, 95; shading, 22–23; shock response, 43, 91, 105; soil augmentation—precautions, 22, 31; watering, 22. See also recommendations for individual species (Section 3)
Pleomele forbesii. See *hala pepe*
Pleomele halapepe. See *hala pepe*
Pleomele hawaiiensis. See *hala pepe*
Plumbago zeylanica. See *ʻilieʻe*
potting, 16, 17–20; post transplant care, 19–20, 43, 50, 66, 72, 77, 100, 113, 132; shock response, 43, 91, 100, 105; soil mixtures, 17–18; soil temperature, 19–20; transplanting technique, 18–19, 100; watering, 19–20, 72, 93. See also recommendations for individual species (Section 3)
Pouteria sandwicensis. See *ʻālaʻa*
powdery mildew, 32, 141
prickly poppy. See *pua kala*
Pritchardia aylmer-robinsonii. See *loulu*
Pritchardia glabrata. See *loulu*
Pritchardia napaliensis. See *loulu*
Pritchardia remota. See *loulu*
Psychotria hathewayi. See *kōpiko*
Psydrax odorata. See *alaheʻe*
pua kala, 2, 3, **134–136**
Pycnontus cafer. See bulbul

Pycnontus jocosus. See bulbul
Pythium ultimum. See damping-off

rat, 3, 5, 104, 137
Rauvolfia sandwicensis. See hao
reserpine, 60
Reynoldsia sandwicensis. See *'ohe makai*
Rhizoctonia solani. See damping-off
Rock, Joseph, 42, 62, 112
rooting hormone, 16, 52, 90
roundworms. *See* nematodes

sandalwood, 64–68, 114
Santalum ellipticum. See *'iliahi*
Santalum freycinetianum. See *'iliahi*
Santalum paniculatum. See *'iliahi*
Sapindus oahuensis. See *āulu*
Sapindus saponaria. See *mānele*
Scaevola coriacea. See *naupaka*
Scaevola gaudichaudii. See *naupaka*
Scaevola sericea. See *naupaka*
scale insect, 26, 37, 38, 48, 52, 57, 59, 61, 63, 68, 83, 88, 97, 101, 107, 119, 130, 132
scarification, 13, 36, 49, 85, 92, 97, 102, 107, 120, 138, 140; hot water, 13, 36, 88
seedlings, 15–16, 17–20. *See also descriptions and recommendations for handling in accounts of individual species (Section 3)*
seeds: descriptions. *See accounts of individual species (Section 3);* diseases and pests, 12, 32, 50, 51, 54, 58–59, 72, 74; dispersal, 1, 62, 85; handling, 12–15, 54, 63, 104, 127; hormone treatment, 65–66, 125; planting, 14–15, 92, 132; scarification, 12–13, 102, 109, 138; sprouting, 14–16; storage, 12, 123
slugs, 23, 30, 37, 43, 46, 66, 77, 98, 104, 121, 130, 132
snails, 30, 46, 66, 77, 104, 121, 130, 132
soapberry tree, 54
soil, 4–5, 19–20, 21, 33; artificial types, 14, 16; augmentation, 21–22; contamination, 31–32; fertilizing technique, 20, 22; geochemical type, 22, 36, 38, 67, 101, 114; microbial plant symbionts in, 80, 109; moisture, 22, 72, 88, 130; mulching, 22; potting mixture, 17–18; temperature considerations, 20, 72, 109
Solenopsis geminata. See ant, fire

Sophonia rufofascia. See leafhopper, two-spotted
Sophora chrysophylla. See *māmane*
spittle bug, 126
State of Hawai'i, DLNR, 9
State of Hawai'i endangered species law, 9–10, 122
stem-borer, 29–30, 38–39, 80, 105, 137, 139; black, or coffee twig (stem) borer, 30, 72, 75, 81, 114
stem chewing pests, 28, 39–31. *See also accounts for individual species of plants (Section 3)*
stinkbug: black, 28, 57, 121, 141; green, 28
sucking pests, 26–29. *See also* aphid, scale insect, whitefly
sulfur (for disease prevention and treatment), 18, 32, 36, 66, 71, 74

thrips, 109
ti, 58, 131
toad, 16
toxicity, 2, 45
transplanting. *See* potting and planting out
treehopper, 28–29, 112

uhiuhi, 4, 5, 16, 19, 29, 30, 31, 100, **136–139**
'ūlei, 105
utricle, 38, 40

Vaccinium. See *'ōhelo*
Vanduzeea segmentata. See treehopper
Vigna marina. See *nanea*
vitamin B-1, 43

watering, 16, 18, 19–20, 22. *See also recommendations in accounts of individual species (Section 3)*
wauke, 91, 121, 137
weevil, small gray, 25–26, 38, 41, 55, 68, 80, 101
whitefly, 48, 55, 80, 88, 91, 103, 111
Wikstroemia oahuensis. See *'ākia*
Wikstroemia phillyreifolia. See *'ākia*
Wikstroemia sandwicensis. See *'ākia*
Wikstroemia uva-ursi. See *'ākia*
wiliwili, 2, 4, 22, 24, 28, 29, 31, 32, 77, 80, 86, 98, 121, 138, **139–141**
wireworms. *See* nematode

Xylosandrus compactus. See stem-borer, black
Xylosma hawaiiense. See *maua*